THE

ELEMENTS OF ASTRONOMY;

OR

THE WORLD AS IT IS,

AND

AS IT APPEARS.

BY THE AUTHOR OF "THEORY OF TEACHING," "EDWARD'S FIRST LESSONS IN GRAMMAR," ETC.

BOSTON:
PUBLISHED BY CROCKER AND BREWSTER,
47, Washington-street.
1850.

PREFACE.

THE apology for the present treatise on Astronomy is based on the nature of the subject; it is one which requires to be presented to the student from more than one point of view. He should learn from the original observer, the profound generalizer, the investigator of cause and effect in detail; and to prepare him for such studies, he needs the book of one who knows from experience his requirements and his capacity.

As such a book, "The Elements of Astronomy, or the World as it is, and as it appears," is offered by a teacher to the teaching and studying public. Had the writer aimed only to excite an interest in the subject, it would have been shorter and more attractive; but it is intended, likewise, to exercise the student's memory, reason, and imagination. The details introduced for this purpose serve also to keep each truth before the mind some time; they present it in different lights, and secure its being perceived by each pupil fully and in all its bearings.

This book has gradually grown out of lessons given orally during many years of teaching. These were written out for the author's own use, not for publication. Originality was not sought for, and all explanations and illustrations which could be of service to the pupils were adopted. As time passed on, and no book appeared precisely suited to the wants of these pupils, or of High Schools in general, the author began to entertain the idea that these lessons might in some measure supply the want so extensively felt. In this hope such completeness has been given to the work as a very limited leisure would allow. It has been revised in manuscript by George P. Bond, Esq., of the Cambridge Observatory, to whom the author is also indebted for superintending its passage through the press.

Boston, 1850.

CONTENTS.

ELEMENTS OF ASTRONOMY.

CHAPTER I.

DEFINITIONS.

The Sphere. Spherical Distances and Angles. The Spheroid. Earth's Axis and Equator. Terrestrial Latitude and Longitude. The Sphere of the Heavens, Zenith and Horizon. Celestial Pole and Equator. The Ecliptic. The Zodiac. Right Ascension and Declination. Celestial Longitude and Latitude. Parallax. Terms defining the Orbit of a Planet. Sidereal and Apparent Time.

§ **1.** A sphere is a solid terminated by a curved surface, all the points of which are equally distant from a point within called the centre.

The radius of a sphere is a straight line drawn from the centre to a point in the surface; the diameter, or axis, is a line passing through the centre, and terminated each way by the surface.

All the radii of a sphere are equal; all the diameters are also equal, and double the radius.

§ **2.** Every section of a sphere made by a plane is a circle.

If a sphere is cut by a plane which passes through the centre, the section is called a great circle of the sphere; the radius of such a section being the greatest possible, the same, namely, with the radius of the sphere.

From this definition it is evident that a great circle may be made to pass through any two points in the surface of a sphere; and that, if the two points be not opposite extremities of a diameter, only one great circle can be made to pass through them, for its plane must pass through the centre of the sphere, and only one plane can be made to pass through three points which are not in the same straight line. But through the two extremities of a diameter, any number of great circles may be made to pass, for they are in the same straight line with the centre of the sphere.

§ **3.** A diameter of a sphere, perpendicular to the plane of any great circle, is called the axis of that great circle; and the extremities of the axis are called its *poles*.

The angles formed at the centre of the sphere by the plane of a great circle and its axis, are right angles; therefore the pole of a great circle is 90° distant from every point of the circumference of the great circle. The arcs subtending the angles are 90°, and are those of great circles; and all angular distances on the surface of a sphere, to an eye at the centre, are measured by the arcs of *great* circles.

All great circles bisect one another; for all passing through the centre of the sphere, their common section must be a diameter of each, and every diameter bisects a circle.

§ **4.** Secondaries to a great circle are great circles which pass through its poles, and whose planes are therefore perpendicular to its plane. Hence every secondary bisects its great cicle. If it passes through the poles of two great circles, it is perpendicular to each of them and bisects them both. And conversely, if one great circle be perpendicular to two others, it must pass through their poles.

If an eye be in the plane of a circle, that circle appears a straight line; hence in the representation of the surface of a sphere upon a plane, those circles whose planes pass through the eye are represented by straight lines.

The angle formed by the circumferences of two great circles on the surface of a sphere is equal to the angle formed by the planes of those circles; and is measured by the arc of a great circle, intercepted between them, and which is a secondary to each.

§ **5.** If a sphere is cut by a plane which does not pass

through the centre, the section is called a small circle of the sphere; the radius of such a section being less than that of the sphere.

A circle, it is plain, may be made to pass through any three points in the surface of the sphere; and it will be a great or a small circle, according as its plane passes through the centre of the sphere or otherwise.

Parallel circles of a sphere are such as have their planes parallel. Parallel circles have the same axis and poles; for a straight line which is perpendicular to one of two parallel planes is perpendicular to the other likewise. Two parallel circles cannot both of them pass through the centre of the sphere; that is, they cannot both be great circles of the sphere.

If with the intersection of two great circles as a pole, a great circle be described, and also a small circle parallel to it, the arcs of the great and small circles intercepted between the two great circles contain the same number of degrees. And any one of these arcs measures the angle at the pole made by the planes of the two great circles.

The centres of parallel circles lie in the diameter perpendicular to their planes.

§ **6.** Either pole of a circle of the sphere is equally distant from all points in the circumference of that circle; whether the direct or spherical distance be understood.

Hence any circle of a sphere may be conceived to be described from either of its poles as a centre with the spherical distance of that pole as a radius. For if this distance be carried round the pole, its extremity will lie in the circumference of the circle.

The distances of any circle from its two poles are, together, equal to a semi circumference.

A great circle is equally distant from its two poles; but this is not the case with a small circle.

Equal circles of the sphere have equal polar distances, and conversely.

The polar distances of any circle of the sphere are the spherical arcs which join any point in the circumference with the two poles of the circle. By the polar distance (singly) the lesser of these two arcs, or distance from the nearer pole, is generally to be understood.

§ **7.** Any portion of the circumference of a great circle is called a spherical arc.

Two points are said to be joined on the surface of the sphere when the spherical arc between them is described; and this arc is called the spherical distance of the two points, in order to distinguish it from their direct distance, which is the straight line which joins them. The spherical distance of opposite extremities of a diameter of the sphere is evidently half the circumference of a great circle; but the spherical distance of any other two points is less than a semi-circumference, being always the lesser of the two arcs into which they divide the great circle which passes through them.

§ **8.** If the arcs of two great circles meet in one point, they are said to form at that point a spherical angle.

A spherical angle is greater or less according to the opening between its containing arcs.

Every spherical angle is measured by the plane angle, which measures the inclination of the planes of the containing arcs.

When one spherical arc, standing upon another, makes the adjacent spherical angles equal one to another, each of them is called a spherical right angle, and the arc which stands upon the other is said to be perpendicular, or at right angles to it.

The terms acute and obtuse are likewise applied to spherical angles.

Circles are thus said to make right, acute, or obtuse angles with one another. And we may measure this angle by the spherical angle on the surface made at the points where the circles intersect, or by the angles made by the planes of the intersecting circles, or by the angles made by the tangents of the two circles at the point of intersection. It will be the same in whichever way we measure it.

If we draw from the pole of a great circle to any point in its circumference a spherical arc, this arc is a quadrant, or 90° of a great circle, and is at right angles to the circumference. For since the pole is always 90° from the circumference, the axis and plane of the circle have an arc of 90° for the measure of their angle, therefore they make a right angle.

If there be two equal and parallel small circles, and a great circle meets one of them in any point, it will meet the other in the opposite extremity of the diameter which passes through that point.

If a great circle cuts one of two equal and parallel small circles, it will cut the other likewise.

§ **9.** In order to compare together different arcs and angles, every circumference of a circle is supposed to be divided into 360 equal arcs, called degrees, and marked thus, (°). For instance, 60° is read 60 degrees.

Each degree is divided into 60 equal parts called minutes, and marked (′).

Each minute is divided into 60 equal parts called seconds, and marked (″).

§ **10.** As all circumferences, whether great or small, are divided into the same number of parts, it follows that a degree which is thus made the *unit* of arcs is not a fixed value, but varies for every different circle. It merely expresses the ratio of an arc, namely, $\frac{1}{360}$, to the whole circumference of which it is a part, and not to any other.

An angle has a fixed value altogether independent of the radius of the arc by which it is measured. But whatever radius we give the arc, the arc will always have the same proportion to its circle, and this proportion gives the same number of degrees for the measure of the angle.

If we make in the edge of a ruler five notches, and turn the ruler round one of its ends as a centre, making the ruler a radius, and describing a circle with its outer end, the five notches will have described five circles, and the ruler has made four right angles; each of these angles having 90° for its measure.

§ **11.** A tangent to a circle is a line which has only one point in common with the circle, and is perpendicular to a radius of the circle drawn to the point of contact.

A tangent to a sphere has only one point in common with the sphere.

A plane is tangent to a sphere when it touches its surface only at one point. Owing to the form of a sphere, a plane surface can touch its surface only at one point, unless it cuts the sphere.

§ **12.** If we suppose a sphere to be flattened at the poles, we shall have the solid called an oblate spheroid. Its shortest diameter will be through its poles ; its longest diameter will lie in the plane at right angles to its axis. All other diameters will be longer than the former of these, and shorter than the latter. Only one section of it will be a great circle,—the one 90° from its poles. There may be many small circles parallel to this.

If we suppose the sphere to be lengthened out at the poles, it will be a prolate spheroid.

We may suppose a sphere to be generated by the revolution of a circle about its diameter.

An oblate spheroid is formed by the revolution of an ellipse about the shorter axis. A prolate spheroid is formed by the revolution of an ellipse about its longer axis.

§ **13.** The axis of the earth is that diameter about which it rotates with a uniform motion from west to east. The extremities of this diameter or the points where it meets the earth's surface are called the poles of the earth.

The terrestrial equator is a great circle on the earth's surface equidistant from its poles, dividing it into two hemispheres,—a northern and a southern. The plane of the equator is therefore a plane perpendicular to the earth's axis, and passing through its centre.

The terrestrial meridian of a place on the earth's surface is a great circle passing through both the poles and through the place.

§ **14.** The latitude of a place on the earth's surface is its angular distance from the equator. This angle lies at the centre of the earth, but is measured on the meridian of the place. It is reckoned in degrees, minutes, and seconds, northward or southward, according as the place lies.

Parallels of latitude are small circles on the earth's surface parallel to the equator. Every point in such a circle has the same latitude.

The longitude of a place on the earth's surface is the angle made by its meridian with the meridian of some place selected as a point to reckon from. Greenwich, in England, is usually taken for this point, reckoning 180° West and 180° East from Greenwich.

Longitude is also reckoned in time. For as the earth rotates 360° in twenty-four hours, it follows that it rotates 15° in one hour. In this case it is reckoned westward all round the globe. We need only divide the number of degrees of longitude by fifteen and we have the number of hours.

§ **15.** The sphere of the heavens is an imaginary concave sphere of infinite radius, having the centre of the earth, or what comes to the same thing, the eye of any spectator on the earth's surface, for its centre. Every point in this sphere may be regarded as the vanishing point of all the lines parallel to that radius of the sphere which passes through it, seen in perspective from the earth. Every great circle on it is the vanishing line of a system of planes parallel to the plane which passes through it and through the spectator's eye.

§ **16.** The zenith is the point of the sphere of the heavens vertically above the spectator; the nadir, the point 180° distant under his feet. They are therefore the vanishing points of all lines parallel to the direction of a plumb line at his station. The plumb line itself is, at every point of the earth, perpendicular to its spherical surface. At no two stations, therefore, can the actual directions of two plumb lines be regarded as mathematically parallel; they converge towards the centre of the earth. But for very small intervals, (as in the area of a building, or in the same town,) the difference from exact parallelism is so small that it may be practically disregarded. An interval of a mile gives to plumb lines a convergence of about a minute.

§ **17.** The zenith and nadir are the poles of the celestial horizon; that is, they are points 90° distant from every point in it. The celestial horizon is the vanishing line of a system of planes parallel to the sensible and the rational horizon.

The sensible horizon is the actual horizon of the spectator. If we suppose a plane to be extended from a point on the earth's surface to the sphere of the heavens, and another plane parallel to the former to be extended from the earth's centre to the sphere of the heavens, these two planes will cut the sphere in the same line. The first of these planes is called the sensible, the latter the rational horizon, and both coincide with the celestial horizon.

Vertical circles of the sphere are great circles passing through the zenith and nadir, or great circles perpendicular to the horizon. On these are measured the altitudes of objects above the horizon,—the complements to which are their zenith distances.

§ **18.** The poles of the heavens are the points of the sphere to which the earth's axis is directed; or the vanishing points of all lines parallel to the earth's axis. The star nearest to each celestial pole is called the pole star.

The celestial equator, or equinoxial, is a great circle of the heavens marked out by the indefinite extension of the plane of the terrestrial equator, and is the vanishing line of all planes parallel to it.

§ **19.** When the plane of the terrestrial meridian of a spectator is prolonged to the sphere of the heavens, it marks out the celestial meridian of a spectator stationed at that place. The intersection of the spectator's meridian with his horizon makes its north and south points.

The vertical circle which cuts the meridian of any place at right angles, is called the prime vertical; and the points where it cuts the horizon are called the east and west points. Hence the east and west points are 90° distant from the north and south points. These four are called the cardinal points.

§ **20.** Azimuth is the angular distance of a celestial object from the north or south point of the horizon, referred to the horizon by a vertical circle; or it is the angle comprised between two vertical planes; one passing through the elevated pole, the other through the object. The altitude and azimuth of an object being known its place in the visible heavens is determined.

§ **21.** The ecliptic is that great circle in the heavens which the earth really describes, and which the sun appears to describe in the course of the year.

The ecliptic and the equator, being great circles, must bisect each other, and their inclination is called the obliquity of the ecliptic. The points where they intersect are called the equinoxial points. The times when the sun comes to these points, and the points themselves, are called the equinoxes, because the day and night are then equal all over the world.

§ **22.** The ecliptic is divided into twelve equal parts called signs:—Aries, Taurus, Gemini, Cancer, Leo, Virgo, Libra, Scorpio, Sagittarius, Capricornus, Aquarius, Pisces. The order of these is according to the real motion of the earth, and the apparent motion of the sun. The first point of Aries coincides with the position of the sun at the vernal equinox, the first point of Libra with the position of the sun at the autumnal equinox.

The first six signs are called northern, lying on the north side of the equator; and the last six are called southern, lying on the south side.

When the motion of the heavenly bodies is according to the order of signs it is called direct. When it is in a contrary direction it is called retrograde. The real motion of all the planets is according to the order of signs, but their apparent motion is sometimes in an opposite direction.

The zodiac is the zone extending on each side of the ecliptic, within which the motion of the planets, and the moon, and the apparent motion of the sun, are performed.

§ **23.** When a body is referred to the equinoxial, its spherical distance from that is called its declination.

The arc in the heavens which corresponds to terrestrial longitude is called an arc of right ascension. As terrestrial longitudes are reckoned from a determinate point on the equator, so right ascensions require some known point in the equinoxial as the commencement of their reckoning, or their zero point. Some hour-circle or meridian, of the celestial sphere, must be taken for their starting point. The most obvious, indeed, the only point naturally marked on the equinoxial, is the one where the ecliptic intersects it, and though this is not exactly stationary, it has been adopted as the zero point. It serves also as a zero point for reckoning on the ecliptic. The right ascensions are always reckoned eastward from the equinox in degrees or in hours.

The longitude of a star is an arc of the ecliptic intercepted between the first point of Aries, and a secondary to the ecliptic passing through the star. If the body be in our system, and seen from the sun, it is called the heliocentric longitude ; if seen from the earth, it is called the geocentric longitude ; the body being in each case referred perpendicularly to the ecliptic by a plane passing through the eye.

The latitude of a star is its angular distance from the ecliptic, measured upon a secondary to the ecliptic drawn through the star. If the body be in our system, its angular distance from the ecliptic seen from the earth, is called the geocentric latitude ; but if seen from the sun, it is called the heliocentric latitude.

§ **24.** The tropics are two parallels of declination, touching the ecliptic. One, touching it at the beginning of Cancer, is called the tropic of Cancer ; and the other, touching it at the beginning of Capricorn, is called the tropic of Capricorn. The tropics are so called from a Greek word signifying *to turn*, because the sun appears to turn there and descend or ascend in the heavens. The two points where the tropics touch the ecliptic are called the solstitial points, because the sun appears to linger or stand still there. The times when the sun is seen in these points are called the solstices.

The colures are two secondaries to the celestial equator ; one, passing through the equinoctial points, is called the equinoctial colure ; the other, passing through the solstitial points, is called the solstitial colure.

The arctic and antarctic circles, (so named from Arctos, the bear, and anti-Arctos, opposite the bear,) are two parallels of declination ; the former about the north, and the latter about the south pole, the distances of which from the two poles are equal to the distances of the tropics from the equator. These are also called the polar circles.

§ **25.** The altitude of the elevated pole being equal to the spectator's geographical latitude, it follows that all the stars whose polar distance does not exceed his latitude are perpetually visible. A circle described round the elevated pole, with a radius equal to the altitude, is called a circle of perpetual apparition. A circle of equal size round the depressed pole is called a circle of perpetual occultation, because the stars within it are never visible.

All celestial objects within the circle of perpetual apparition come twice on the meridian above the horizon in every diurnal revolution; once above and once below the pole. These are called their upper and lower culminations.

All the heavenly bodies culminate, (i. e. come to their greatest altitudes,) on the meridian, which is, therefore, the best situation to observe them, as they are there least confused by the inequalities and vapors of the atmosphere, as well as least displaced by refraction.

§ **26.** A body is in conjunction with the sun when it has the same longitude; in opposition when the difference in their longitudes is 180°, and in quadrature when the difference in their longitudes is 90°.

Syzygy is either conjunction or opposition.

The elongation of a body from the sun is its angular distance from the sun when seen from the earth.

The diurnal parallax is the difference between the apparent places of bodies as seen from the centre and from the surface of the earth.

The annual parallax is the difference between the apparent place of a body when seen from opposite points of the earth's orbit.

§ **27.** The points where the orbits of the planets cut the ecliptic, and where the orbits of the secondaries cut the orbits of their primaries, are called their nodes. That node is called ascending where the planet passes from the south to the north side of the ecliptic; the other is called the descending node. The line which joins the nodes is called the line of the nodes.

Aphelion is that point in the orbit of a planet which is farthest from the sun.

Perihelion is that point in the orbit of a planet which is nearest to the sun.

Apogee is that point of the moon's orbit which is farthest from the earth.

Perigee is that point of the moon's orbit which is nearest to the earth.

The apsis of an orbit is either its aphelion or perihelion, apogee or perigee ; and the line which joins the apsides is called the line of the apsides.

The true anomaly of a planet is its angular distance at any time from its aphelion or apogee. The mean anomaly is the angular distance from the same point at the same time, if it had moved uniformly with its mean angular velocity.

The equation of the centre is the difference between the true and the mean anomaly.

The mean place of a body is the place where a body (not moving with a uniform angular velocity about the central body) would have been, if it had moved with its mean angular velocity. The true place of a body is the place where the body actually is at the time.

The corrections applied to the mean place of a body in order to get its true place are called equations.

§ **28.** Apparent noon is the time when the sun comes to the meridian. True or mean noon is twelve o'clock by a clock adjusted to go twenty-four hours in a mean solar day. The equation of time at noon is the interval between true and apparent noon.

Sidereal time is reckoned by the apparent diurnal motion of the stars, or rather of that point in the equinoctial from which right ascensions are reckoned. The interval between any two successive returns of this to the meridian is called a sidereal day, and is divided into twenty-four sidereal hours.

CHAPTER II.

LIGHT AND THE TELESCOPE.

Nature and properties of Light. Velocity of Light. Transmission and Reflection. Lenses, and the Refracting Telescope. Magnifying Power. Fields of View. Distinction between the different powers of a Telescope. Reflecting Telescope. Achromatic Telescope.

§ **29.** Philosophers are not agreed about the nature of light. Some maintain that it is an emanation from luminous bodies; others suppose it to be produced like sound, by the undulations of a subtile fluid diffused throughout space. Many facts favor this last theory. Rays of light cross in every direction without interfering; and we can suppose this more easily of waves than of actual particles. If light is a substance, what becomes of it? where is all that has been emitted since the beginning of the world? If it is merely a motion of particles, it may cease like any other motion. As the undulatory theory has much in its favor, we shall adopt it and consider light as that vibration of an unknown fluid which causes to our eye the sensation of sight. We cannot attain perfect certainty as to the nature of light, but we can learn its laws, and reason concerning them without this certainty.

§ **30.** Bodies are divided into luminous and non-luminous. All light comes originally from some self-luminous body. It is emitted from every point of such a body, and in every direction in which the point is visible. A ray is a single line of light; a pencil of rays is a collection proceeding from any one point of such a body. These rays move through a uniform medium in straight lines, entirely independent one of another. We know that rays move in straight lines, for no opaque body can screen a luminous point from us unless interposed in a straight line between us and it. Light moves with a velocity of 192,500 miles in a second. It travels from the sun to the earth in eight minutes; a distance which the earth, moving nineteen miles a second, will take two months to pass through.

It moves through a space equal to the circumference of our globe in the eighth part of a second, a flight which the swiftest bird could not perform in less than three weeks. Light diminishes in intensity according to the squares of the distance. At a certain distance let it be diffused over a certain space and have a certain intensity, at twice the distance it is diffused over four times the space, and has but one fourth the intensity.

§ **31.** Luminous bodies, probably from being themselves in agitation, cause vibrations in some medium which excites our optic nerve, and makes us conscious of the object at which the vibration began. Hearing, which seems to be a coarser sense, is excited in the same manner. But we know that the air, wood, ice, rock, or almost any substance, may serve as a medium for the vibrations which affect the auditory nerve. While we cannot detect the medium which transmits vibrations to the optic nerve, as the transmission is so much more rapid than in case of sounds, and as we see to far greater distances than we hear, we cannot but feel it probable that the medium is more thin and subtile, and that no other can be substituted for it. This medium must extend to the stars, otherwise no vibrations from them could reach us; it must exist in interstices of transparent bodies, such as air, water, and glass. Other facts also make it probable that regions beyond our earth are occupied by a thin ether. The comets, whose frequent returns enable us to calculate their periodic times, seem to be drawn nearer and nearer to the sun, their times are diminishing. If so, they must meet some resistance which gradually diminishes their projectile power, and this resistance must come from an invisible thin fluid. A fluid too thin to act sensibly upon the planets, but dense enough to derange the lighter comets, and to transmit light.

§ **32.** Non-luminous bodies are either opaque or transparent. When light falls on an opaque body, a part of it is reflected, a part enters the body and is lost. When the body is transparent, the greater part is transmitted, and but little is reflected. Both reflection and transmission follow particular laws, which must be understood previous to studying astronomy.

Bodies are referred to the direction from which the ray by which we see them comes to our eyes. We cannot know, except by reasoning, what changes this direction may have undergone.

§ **33.** Reflected light serves as well as original light to show us the direction of bodies. We see the moon, and the clouds, we see objects around us by light which may have been many times reflected. Reflected light is however weak, because a great portion of that which falls on bodies is usually absorbed. There is one condition necessary to our seeing a body by reflected light. The angle by which the light passes to us is always equal to that by which it strikes upon the reflecting body. The angle of reflection is equal to the angle of incidence. We must, therefore, be in such a position that the ray reflected at this angle shall enter our eye, or we shall not see the body.

§ **34.** When light passes obliquely from a medium to one more dense, it is bent at the surface of the denser medium into a line more nearly approaching a perpendicular. It passes without bending through the medium, and if it again passes into the thinner medium, it is turned away from the perpendicular by an angle equal to that at which it entered. The more obliquely it enters the more it is turned aside ; the sine of the angle of incidence always bears the same ratio to the sine of the angle of refraction, when the same media are compared. When light passes perpendicularly from one medium to another, it suffers no refraction.

By interposing between the eye and the object a material of higher refractive power than air, and making it of such a form that the refraction at one or both surfaces shall bring the rays to a focus behind it, we bring the image of the object nearer to the eye, and thus increase its apparent size, brightness, and rapidity of motion.

§ **35.** The material used for the lenses of telescopes is glass. The lenses used are chiefly the plano-convex, the double convex and the double concave. Lenses of the same material converge the rays in proportion to the convexity of their form. A small portion of a large sphere refracts to a longer focus than a large portion of a small sphere.

The simplest telescope consists of a plano-convex lens, whose focal length exceeds six inches, placed at one end of a tube whose length must be six inches greater than the focal length of the lens. Six inches being the distance at which an eye sees small objects with ease. An eye placed at the other end of the tube will see an inverted image of distant objects magnified in proportion to the focal length of the lens. If the lens has a focal length of from ten to twelve feet, the magnifying power will be from twenty to twenty-four; to a very shortsighted person, who sees objects distinctly at three inches, the magnifying power would be from forty to forty-eight.

§ **36.** It is well known that the same object brought nearer to the eye subtends a larger angle, and therefore appears larger than when more distant. It is also true that a small body, very near, subtends a larger angle, and therefore appears larger than a distant object of greater size.

Let us examine the effects of placing convex lenses at different distances between the eye and an object. We will suppose a man to be 100 feet from the eye. We will place a convex glass of twenty-five feet focal distance half way between the man and the eye. The inverted image of the man will be formed fifty feet behind the lens, of the same size as the object. This image can be seen by the eye placed six inches behind it with great distinctness, and nearly as well as if the man had been brought from the distance of 100 feet to six inches. Thus the man, though seen no larger than life, would be apparently magnified, because his apparent magnitude would be increased in the proportion of 100 feet to six inches. If a lens of shorter focus be placed between them in such a way that the man is twenty feet before the lens, and his image eighty feet behind it, the size of the image would be four times that of the object. The man would be magnified four times directly by the lens, and 200 times by being brought nearer to the eye; in the whole 800 times.

§ **37.** When the focal length of the lens is quite inconsiderable compared with the distance of the object, as it is in viewing the heavenly bodies, the rule becomes this:—

Divide the focal length of the lens by the distance at which the eye looks at the image. If we use a lens with a focal distance of ten feet, on account of the sun's being 95,000,000 miles distant, the image of the sun formed in the focus will be only $\frac{10 \text{ feet}}{95,000,000 \text{ miles}}$ of the true size of the sun's disc. But this image is viewed at a distance of six inches instead of 95,000,000, therefore it is magnified $\left.\frac{95,000,000 \text{ miles} \times 10 \text{ feet}}{95,000,000 \text{ miles} \times 6 \text{ inches.}}\right\} = 20$ times.

§ **38.** A lens placed in a telescope as we have described it, is called an object glass. It bestows on a telescope the power of bringing the image near, which is evidently the same thing as transporting the observer to the distant regions of space. But there is another way of increasing the apparent magnitude of objects, particularly of those which are within our reach, which is of great importance in optics. Rays from each point of a distant object enter the eye in parallel lines, and the object is distinctly seen. If we bring an object very near, so as to give it great apparent magnitude, it becomes indistinct; because the rays diverge from each point so much that the eye cannot bring them to a focus. But if we can make the rays from it enter the eye nearly parallel, it will be distinctly seen. Since parallel rays are brought to a focus by a lens, if we place an object or its image in the focus of the lens, the rays will come out from it parallel. We shall see the object very distinctly, and it will be magnified in the proportion of six inches, at which we see small objects most distinctly, to its present short distance from the eye. But this short distance is equal to the focal length of the lens, so that the magnifying power produced by the lens is equal to six inches divided by the focal length of the lens.

Such a lens is called the eyepiece of the telescope. These two lenses together constitute a simple form of the astronomical refracting telescope.

§ **39.** Let us consider the effect of each of these glasses upon the apparent motion and brightness of objects seen through them. Distant motions appear slow, because the space passed through subtends so small an angle at the eye. The moon wheels round the earth 2,000 miles an hour; yet owing to her distance from it her motion is not visible to the naked eye. The same space described by two objects, one a hundred times more distant than the other, would seem a hundred times smaller in the former case than in the latter. The object-glass, since it brings objects nearer, must therefore increase their apparent motion in the same ratio.

§ **40.** A very rapid motion may be imperceptible, if the distance of the moving body be sufficiently great. For, the greater the distance, the smaller the angle under which the motion appears to the eye; and if the arc which a body describes in an hour does not subtend an angle of more than 20°, its motion is imperceptible. The greatest apparent motion in the heavens does not exceed 15° in an hour, being that produced by the rotation of the earth. The greatest central motion is that of the moon and does not exceed about 13° a day.

The apparent motion of heavenly bodies caused by the rotation of the earth is magnified by the object glass. The field of view appearing larger than it really is, bodies appear to pass through it in the same time more swiftly than they would to the naked eye.

The eye piece also increases apparent motion in proportion to its power of magnifying. As in the solar microscope where the motions of animalcules appear prodigiously swift.

Since the object and the eye glasses affect motion similarly, we shall in the following calculations consider not the power of each separately, but the whole power of the telescope.

§ **41.** Let a person direct the tubes of a telescope (without the lenses) to any celestial object, and there fix them; he will soon find that in a short space of time the object will have removed from before the mouth of the tube. Now this motion of the celestial bodies, which is only *apparent*, arises from the rotation of the earth upon her axis; and the quantity of this motion may be determined with facility thus: the earth is known to revolve once about her axis in twenty-four hours, and as every circle is supposed to be divided into 360 equal parts or degrees, the apparent time any celestial body takes to describe one degree will be found by dividing the 24 hours by 360, which gives us four minutes as the time an object in the celestial equator would take to pass the mouth of the tube if it only takes in one degree of the heavens.

In ordinary telescopes the field of view is about half a degree, and an equatorial star will cross it in two minutes. The moon's diameter is 30′, it will therefore fill the field of such a telescope, and the whole moon, from the time one edge enters till the other leaves, will require four minutes. If now we suppose the glasses to be put in the tube, the magnifying power will cause all bodies nearer than the fixed stars to look larger, and to appear to move faster, but the size of the tube remaining the same, no larger portion of the heavens from side to side can be seen through it than before. On the contrary, as a general rule, the higher the magnifying power the smaller are the linear dimensions of the field of view, while the apparent diameter and motion of objects is increased.

§ **42.** For convenience of calculation the linear dimensions have been used. But in truth when a surface is magnified it is so in all directions, and the increase of surface will be according to the square of the linear increase. Thus when the moon is brought by the telescope ten times nearer, its disc is one hundred times increased, and the area of the field of view one hundred times diminished. In the case of the fixed stars, while the telescope increases their brilliancy, no magnifying power that has yet been applied has shown their true discs, owing to their immense distance from us.

§ **43.** There are three modes in use among astronomers of designating the capacity of a telescope. The first is by its illuminating power, which is proportioned to the area of its object glass; the second by its magnifying power, which may be increased to any extent by changing the lenses of the eye pieces; and the third is by its space-penetrating power, by which is meant such a relation between the illuminating and the magnifying powers that the object viewed shall appear with its natural brightness.

That is, of the brightness with which the object would appear to the naked eye were the observer transported to a point as many times nearer to the object as the magnifying power employed is greater than unity.

In all cases the space-penetrating power of a telescope is limited by the area of its object glass, or of its speculum, if it be a reflector.

§ **44.** On account of the difficulty of obtaining large lenses of a sufficiently pure material, a concave mirror is sometimes used instead of an object glass. The telescope is then called a reflector, as the rays of light are brought to a focus by reflection from the polished surface of the speculum. Large reflecting telescopes are not so well adapted to delicate observations and measurements as to discoveries in physical astronomy, the resolution of nebulæ, and the examination of extremely faint objects. Those of Herschel and Lord Rosse are the most celebrated. In Herschel's forty foot reflector the observer turns his back to the object and looks in at the mouth of the telescope tube, near to the edge of which the image is thrown by a slight inclination of the mirror. Lord Rosse's telescope has a speculum of six feet diameter and a tube of sixty feet.

§ **45.** There is a difficulty both in convex lenses and concave mirrors arising from their shape, called spherical aberration. The rays from the central part of the lens or mirror are brought to a focus later than those from the edges and the rest of the surface. Spherical aberration causes indistinctness of vision by spreading out every mathematical point of the object into a small spot in its picture; these spots by mixing with each other confuse the whole.

It is avoided in mirrors by making the surface not spherical, but of the shape formed by the revolution of a parabolic curve. The rays reflected from the edges of the mirror are, on account of the form of the surface, brought to a focus as soon as those from the central portion. It is destroyed in refracting telescopes by combining a meniscus with a double convex lens, or by giving certain proportions to the figure of a single lens.

§ **46.** Rays from each point of the object spread over the whole lens and should be refracted to a point; but if refraction is unequal in different parts of the lens, rays from a point cannot be converged to a point, but either the more or the less converging will be spread out.

There is also a difficulty in refracting telescopes, owing to the differing refractions of the different colored rays. Violet rays, being the most refrangible, come to a focus sooner than those of any other color. Red rays, being less so, will have a longer focus. Any image formed in the focus of violet rays will be violet colored, and images may be formed between this and the focus of red rays of each color of the spectrum. Therefore no white image can be formed, for there is no one spot in which all the colored rays will be present.

§ **47.** If the dispersive powers of different media were in proportion to their refractive powers, it would be impossible to correct this chromatic aberration. But fortunately different media produce spectra of different lengths when the mean refraction is the same. Let us compare two prisms, one of crown glass, the other of flint glass, with such a refracting angle that the light shall enter and quit them at equal angles, the mean ray of each will have the same refraction. But the spectrum produced by the flint glass will be longer than that produced by the crown prism. Thus flint glass is said to have a greater dispersive power than crown glass, because at the same angle of mean refraction it separates farther the extreme rays of the spectrum. Diamond has a refraction nearly three times that of glass, while its dispersive power is less than that of glass.

§ **48.** Now if different media produce different bands of color with the same focal length, it follows that they may produce equal bands of color with different focal lengths. A concave lens may be used with a convex lens of equal dispersive power, but a higher refractive power, and the excess of the refractive power, will be the available power of the lens, and white light will be refracted to the focus. Such a lens is called an achromatic lens, and the image formed by it would be perfect were there not in equal spectra formed by different media a difference which prevents their entirely neutralising one another's refraction. The bands of the same color in the two spectra are not of equal breadth, and therefore the images seen through such a lens are bordered on one side with a purple, on the other with a green fringe.

A telescope which is free from dispersion is called achromatic ; one which is free from aberration also is called aplanatic, or free from all errors.

Reflecting telescopes are perfectly free from color. For compound light is reflected, though not refracted, entire, all the colors following the same law of equal angles of incidence and reflection.

CHAPTER III.

ASTRONOMICAL INSTRUMENTS.

Difficulties in the construction of Telescopes. Telescope Stands. Transit Instrument. Graduated Circle and Vernier. Mural Circle. Polar and Horizontal Point. Transit Circle. The Equatorial. The Altitude and Azimuth Instrument. Theodolite. Sextant. Difficulties in Observing. Personal Equation. Lord Rosse's Reflectors.

§ **49.** Though the theory of the construction of telescopes is attended with many difficulties, those which occur in practice are as numerous, perhaps some of them are insuperable. For a reflector, a perfectly uniform metal is required, free from all microscopic pores, not liable to tarnish, not so hard as to be incapable of taking a good figure and an exquisite polish, not so soft as to be easily scratched. Various compositions of metals are employed, consisting chiefly of copper and tin, with a little zinc, arsenic or silver. After casting they must be ground and polished with the utmost care. Lenses also require the greatest nicety in composition, casting, grinding, polishing, and centering. To ascertain if the shape is perfect, an opaque back is placed behind the lens, the lens is made to revolve, and a lighted candle is brought before it, whose reflected image is attentively watched. If this image has any motion, the lens is not perfect in its adjustment.

§ **50.** After an instrument is completed the next desideratum is a steady and immovable stand, free from vibration. The instrument should be supported at both ends to give steadiness, and to prevent its being affected by the wind; for every vibration will be increased in the same ratio as the amplification of the instrument, and produce a tremulous or dancing motion in the objects. Thus a superior telescope badly supported may be inferior to a common one on an immovable stand.

The materials of which stands are composed should be capable of transmitting as little vibration as possible; the vibration of a frame of cast iron in one piece, though otherwise perfectly steady, would be sufficient to destroy distinct vision. The difficulty of preventing vibrations in reflecting telescopes greatly impairs their value, as they are more affected by such disturbance than refractors. A telescope, which taken from its stand and placed on a lump of soft clay would enable a person to read a bill placed at a distance of 900 feet, would on its stand make it distinct only at a distance of 650 feet, although no tremor would be discerned on the stand.

§ **51.** A difficulty in placing an instrument arises from the absence of natural indications, other than those afforded by astronomical observations themselves, whether an instrument has or has not its true position with respect to the horizon and its cardinal points, the axis of the earth, or to other principal astronomical lines and circles. For instance, to place a transit instrument correctly, we must know the direction of our meridian, but we must first learn this meridian approximately by observing the shadow cast by the sun at noon.

The transit instrument consists of an astronomical telescope with wires and a micrometer, and is mounted on a nicely formed axis at right angles to itself. This axis remaining always horizontal and directed to the east and west points of the horizon rests at its extremities in two sockets perfectly even, and set in two blocks of stone of a size and weight sufficient to prevent all agitation. The smooth extremities of the axis are capable of nice adjustment by screws, both in a vertical and horizontal direction. By placing a spirit level on the points, the axis can be made perfectly horizontal. Whether the axis lies precisely east and west can only be nicely ascertained by observations made with the instrument itself. When it is perfectly well adjusted the central line of the telescope will not quit the plane of the meridian when the instrument is turned round on its axis. The transit instrument is used to note the passage of bodies over the meridian, to note the right ascension of the fixed stars, the upper and lower culminations of the circumpolar stars, and for various problems in time and longitude.

§ **52.** In the focus of the eye-piece and at right angles to the length of the telescope is placed the system of wires. This consists of one horizontal and five equidistant vertical threads or wires, which always appear in the field of view when properly illuminated, by day by the light of the sky, by night by that of a lamp. The horizontal wire is fixed, the middle vertical wire is brought to bisect the axis of the telescope, and thus to coincide with and represent that portion of the celestial meridian which appears in the field of view. When a star crosses this wire it culminates, or passes the celestial meridian. The instant of this event is noted by a clock or chronometer, an indispensable accompaniment of the transit instrument. For greater precision, the moment of crossing each of the five or seven vertical threads is noted and a mean taken between the times thus obtained, the threads being equidistant; this tends to subdivide and destroy the errors.

An important observation of its correctness consists in reversing the ends of the axis, or turning it east for west. If this be done, and it gives the same results, we may be sure that the line of collimation of the telescope is at right angles to its axis, and marks out in the heavens a great circle.

§ **53.** To measure any small angular distance with a micrometer, as the diameter of a planet, two parallel wires are made to approach to or recede from each other till the body to be measured is exactly inclosed by them. Having accurately measured the planet by the two cross wires, we must next ascertain their distance asunder. The wires are moved by screws. The very slow motion which may be imparted to the end of a screw by a very considerable motion in the power, makes it very useful in the measurement of minute motions and spaces. Suppose a screw cut so as to have fifty threads in an inch, each revolution of the screw will advance its point through the fiftieth part of an inch. Now suppose the head of the screw to be a circle whose diameter is one inch, the circumference of the head will be 3.14 inches. This may easily be divided into a hundred equal parts distinctly visible. If a fixed index be presented to this graduated circumference, the hundredth part of a revolution of the screw may be observed by noting the passage of one division of the head under the index. Since one entire revolution of the head moves the point through the fiftieth of an inch, one division will correspond to the five thousandth part of an inch.

Micrometer threads are made of spiders' webs, India-rubber and glass threads, hair and wires. By thickly coating a fine platina wire with silver and drawing it out as fine as possible, and then dissolving the silver but not the platina, a very fine wire is obtained.

§ **54.** The angular intervals measured by means of the clock and transit instrument, are arcs of the equinoctial intercepted between the hour circles passing through the objects observed. Their measurement is performed by no artificial graduation of circles, but by the help of the earth's diurnal motion, which carries equal arcs of the equinoctial across the meridian, in equal times, at the rate of 15° per sidereal hour.

In all other cases, when angular intervals are to be measured, circles or portions of circles are referred to, others constructed of metal, and mechanically subdivided into equal parts, such as degrees, minutes, &c. The instrument is sometimes movable upon the circle, sometimes both revolve together on an axis concentric with the circle, and forming one piece with it. As the telescope and circle revolve through any angle, the part of the limb of the latter, which by such revolution is carried past the index, will measure the angle described. The index may be a simple pointer, like a clock hand, or a compound microscope, furnished with wires movable by a fine threaded screw, or a vernier, so called from the name of the inventor.

§ **55.** The vernier used in astronomical observations is a small arc of a circle, graduated so as not to correspond with the graduation of the circle, with which it is to be used. For instance, let the large circle be divided into tenths of inches, eleven tenths of an inch are on the vernier divided into ten equal parts. Each division of the vernier then contains $\frac{11}{10}$ of each division of the circle.

When applied to the circle, the first division enables you to measure $\frac{1}{100}$ of an inch, the second division $\frac{2}{100}$, and so on. This is done with much more accuracy than if each tenth of an inch were divided into ten equal parts. Besides only a small arc need be divided, and this can be screwed on wherever it is required. Double verniers and even those more highly divided are used. They are read off by microscopes.

§ **56.** It is no easy thing to divide the circumference of a circle accurately into 360 equal parts, and these again into smaller subdivisions. An angle of one minute occupies on the circumference of a circle of ten inches in radius only about $\frac{1}{350}$ part of an inch, a quantity too small to be certainly dealt with without the use of magnifying glasses; yet one minute is a gross quantity in the astronomical measurement of an angle. With the instruments now employed in observations, a single second is rendered a distinctly visible and appreciable quantity. Now the arc of a circle subtended by one second is less than the 200,000th part of the radius, so that on a circle of six feet in diameter it would occupy no greater linear extent than $\frac{1}{5700}$ part of an inch; a quantity requiring a powerful microscope to be discerned at all. Modern artists, however, carry these divisions to great delicacy. A circle of three and one quarter feet in diameter may now be divided into 10,800 equal parts; and each of these by its accompanying micrometrical apparatus into 1,200 subordinate intervals.

§ **57.** A large graduated circle, such as has been described, is sometimes supported in the plane of the meridian, on a long and powerful axis, and this axis is let into a massive wall or pillar. It is hence called a mural circle. The meridian being at right angles to all the diurnal circles described by the stars, its arc intercepted between any two of them will measure the least distance between them, and will be equal to the difference between their declinations, or to the difference between their meridian altitudes. These differences are then the angular intervals directly measured by the mural circle. But from these it is easy to conclude not only their differences but the altitudes and the declinations themselves ; for the declination of a body is the complement of its distance from the pole. The pole being a point in the meridian might be directly observed on the limb of the circle, if any star stood exactly therein ; and thence the polar distances and the declinations of all the other stars might be at once determined. But this not being the case, a bright star near the pole is selected and observed in its upper and lower culminations ; that is when it passes the meridian above and below the pole. Now as its distance from the pole remains the same, the included arc equals twice the polar distance of the star. The polar point being known, the polar distances become also known.

§ **58.** The polar star, which is very brilliant, and only one and a half degrees from the pole, is usually chosen for this purpose. Both its culminations taking place at great and not very different altitudes, the refractions are nearly equal. Its brightness also allows it to be easily observed in the day time. This star is useful for the adjustment and verification of instruments of almost every description. In the case of a transit, it furnishes a ready means of ascertaining whether the plane of the telescope's motion coincides with the meridian. For since this latter plane bisects its diurnal circle, the eastern and western portion of it require equal times for their description. If, therefore, the upper and lower transits follow at equal intervals of twelve sidereal hours, we may conclude that the plane of the telescope's motion is in the meridian.

The place of the polar point on the limb of the mural circle once determined, becomes an origin, or zero point, from which the polar distances of all objects referred to other points on the same circle are reckoned.

§ **59.** A point on the limb of the mural circle, not less important than the polar point, is the horizontal point, which, being once known, becomes in like manner an origin or zero point, from which altitudes are reckoned. The principle of its determination is nearly the same with that of the polar point. Two points are to be found on the limb, one of which shall be as far below the celestial horizon as the other is above it. For this purpose a star is observed at its culmination, by pointing the telescope directly to it, and again by pointing to the image of the same star reflected in the still, unruffled surface of a fluid in perfect rest. The image is as much depressed beneath the horizon as the star is elevated above it. The point of bisection of the arc which measures their distance is the horizontal point.

§ **60.** A divided circle is sometimes permanently fastened at the axis of a transit instrument, the reading being performed by microscopes fixed on the piers. It serves for the simultaneous determination of the right ascensions and polar distances of objects observed; the time of transit being noted by the clock, and the circle being read off by the microscope. This is called the transit circle.

§ **61.** The transit and mural circle are essentially meridian instruments. But we should possess the means of observing an object not only on the meridian, but at any point in its course, or wherever it may present itself in the heavens. Now a point in the sphere is determined by reference to two great circles, one of which passes through the pole of the other. On the earth the position of a place is known if we know its longitude and latitude; in the heavens if we know its right ascension and declination; in the visible hemisphere if we know its altitude and azimuth.

To observe an object at any one point we must be able to direct the telescope to it. The telescope must therefore be capable of motion in the planes at right angles to each other; and the amount of its angular motion in each must be measured on two circles, whose planes must be parallel to those in which the telescope moves. This is effected by making the axis of one of the circles penetrate that of the other at right angles.

§ **62.** There are but two positions in which such an apparatus can be mounted so as to be of any practical utility in astronomy. The first is when the principal axis is parallel to the earth's axis, and therefore points to the poles of the heavens which are the vanishing points of all one set of parallels; and the perpendicular to this axis circle has the equinoctial for its vanishing circle, and measures by its arcs read off hour angles or differences in right ascension. In this position the apparatus is called an equatorial. It is one of the most convenient instruments for all such observations as require an object to be kept long in view, because being once set upon the object, we can follow it as long as we please by a single motion, by turning it round on its polar axis. In many observations this is an inestimable advantage. To counteract the apparent diurnal motion of the celestial objects, which is continually throwing them out of the field of ordinary telescopes, (a great annoyance, especially when high powers are employed,) a clock-work is attached to the equatorial axis, so constructed as to give to the instrument a quiet and steady sidereal motion, contrary to the motion of the earth, and which by a slight modification may be applied to the solar or lunar motion; but it is generally sufficient when adjusted to a star. The effect of this arrangement is to keep the object for several hours constantly in the centre of the field of view.

§ **63.** The other position for such an apparatus is that in which the principal axis occupies a vertical position, and one circle corresponds to the celestial horizon, and the other to a vertical circle of the heavens.

The angles measured on the former are azimuths or differences in azimuth, and those on the latter zenith distances or altitudes, according as the graduation commences from the upper point of its limb, or from a point 90° distant from it. The vertical position of its vertical circle may be known by a plumb line, which, however the circle be turned round, should always intersect a mark placed near its lower extremity. The north or south point on the horizontal circle, is ascertained by bringing the vertical circle to coincide with the plane of the meridian. If the zero on the horizontal circle is brought into the plane of the meridian, and the telescope is turned to a star east or west of that circle, the azimuth of the star may be read off the horizontal circle at once.

§ **64.** The north or south point on the horizontal circle may likewise be ascertained by the method of equal altitudes. Let a bright star be observed at some distance east of the meridian, by bringing it on the cross wires of the telescope. In this position let the horizontal circle be read off, and the telescope securely clamped on the vertical one. When the star has passed the meridian, and is descending, let it be followed by moving the whole instrument round to the west, without however unclamping the altitude circle until it comes into the field of view, and until, by continuing the horizontal motion, the star and the cross of the wires come once more to coincide. In this position it is evident the star must have the same altitude above the western horizon which it had when first observed above the eastern. At this point let the motion be arrested and the horizontal circle be again read off. The difference of the readings of the horizontal circle will be the arc of azimuth described in the interval, and the north or south point of the horizon will bisect this arc.

An altitude and azimuth circle is particularly useful in investigating the amount and laws of refraction. The paths of stars can be directly traced, and the exact form learned into which their orbits are distorted by refraction.

§ **65.** The theodolite is a modification of the altitude and azimuth instrument. It is devoted to measuring horizontal angles between terrestrial objects.

The sextant is an instrument of great service in nautical astronomy. Its construction is simple, consisting of a graduated arc of 60°, a small telescope, a movable arm with a vernier attached, and two plane mirrors, the one fixed and the other moving with the vernier. It is used in measuring the angular distances of objeets from each other; as of the sun from the horizon, or of the moon from neighboring stars. This little instrument is justly regarded as one of the most useful inventions of modern times, since it is only under its guidance that distant voyages can be successfully accomplished.

§ **66.** Astronomical telescopes have usually a small telescope, called a finder, attached to them, with a magnifying power of not more than ten, and a proportionally large field of view. In its focus two wires cross each other at right angles. When the finder is properly adjusted, an object which has been brought to the intersection of the wires, will be seen in the centre of the field of view of the large telescope.

§ **67.** The most important astronomical instruments have now been described; it remains to say a little about the qualities needed in an observer. Great manual dexterity, accuracy, promptness, and judgment, are required merely to make the instrument perform all of which it is capable. The least inexpertness, defective vision, slowness in seizing the exact instant of the occurrence of a phenomenon, or precipitancy in anticipating it, any one of these destroys the value of observations. The constant care and vigilance of the practical astronomer must therefore be directed to the detection and compensation of errors. He cannot get rid of them, he must therefore allow for them.

A curious fact connected with observation has been lately recognized. No two observers seem to agree precisely in noting the exact instant at which a star crosses the spider's line of a micrometer. Hence arises what has been termed a personal equation or correction, which must be taken into account when the observations of different observers are involved in the same calculation. There is the same difficulty in fixing the exact distance and relative position of two stars, and this cause of error acts differently with different observers, and in different angles of position. Apparently our judgment of parallelism is greatly affected by our attitude, and especially by the difference between looking up and looking down.

§ **68.** A difficulty in observation of small objects arises from the great light spread around by the brighter stars.

Their presence in the field of view is announced by a dawn like that of morning, and the astronomer who would keep his sight so delicate as to perceive small bodies must protect his eyes from this light. For this purpose the larger star is sometimes hidden by a fine needle introduced into the focus of the eye-piece of the micrometer. The needle, which looks like a black bar, is so placed that the star comes into the field of view behind it, and leaves the eye undazzled and able to observe small objects in the vicinity. The large fixed stars, though they have no discs, require as broad a bar to hide them as a planet, owing to their extreme brilliancy.

Except in that part of the sky which is very near the sun even moderately bright stars may be seen at noon-day through the telescopes. Very bright stars may even be discerned without a glass from the bottom of a well or through the shaft of a tall chimney or a mine.

§ **69.** I cannot better close this account of the obstacles which practical astronomers encounter, than by a description of the manner in which they have been triumphantly overcome by Lord Rosse in the construction of his mammoth telescope. Lord Rosse commenced his labors twenty years ago. He began by attempting the improvement of the refracting telescope, but soon gave preference to the reflector. He endeavored to produce the true parabolic speculum which should be free from aberration. The exceeding delicacy needful in producing this form with mathematical accuracy may be judged from the fact, that if two specula of six feet in diameter, the one spherical and the other parabolic, were pressed into contact at the centre, the edges would not diverge from each other more than the thousandth part of an inch. He invented a grinding and polishing machine, and after repeated trials obtained the means of furnishing specula from one to six feet in diameter.

§ **70.** A difficulty no less formidable impeded his operations,—the casting a speculum of sufficient size and strength. After repeated trials he made one of three feet in diameter, cast in sixteen separate portions. By the experience he acquired in making this he became acquainted with the method of casting a large speculum in a single piece. Several tormenting difficulties attended his first efforts. Small air-holes were formed in the metal, and the speculum cracked in cooling. A mould of sand, and subsequently a mould of cast-iron, failed in giving freedom from pores. The desideratum was a kind of mould which should retain the molten metal, and yet allow the air globules to escape. Such was at length discovered, and stamped Lord Rosse's name with celebrity, reducing as it does the casting of specula to a certainty. The contrivance consisted in making the bottom of the mould of layers of hoop iron, bound closely together, with the edges uppermost. The iron conducted the heat away through the bottom so as to cool the metal towards the top, while the interstices between the hoops, though close enough to prevent the metal from running out, were sufficiently open to allow the air to escape.

The first large speculum thus made in a single piece was a round plate of metal, three feet in diameter, nine inches thick, and upwards of a ton in weight. In a few minutes the metal set in a compact form, and while intensely hot, was conveyed by a railway to an annealing oven, a few feet distant from the foundry. The oven was nearly red hot when the speculum was shut up in it, and from this temperature it was allowed many weeks to become gradually cool. It was then ground to the proper parabolic curve, and polished.

§ **71.** It was ascertained in the following manner when the proper parabolic curve was produced. A high tower was erected immediately over the speculum. On a pole at the top of the tower, ninety feet distant from the speculum, the dial-plate of a watch was placed, forming a small round object relieved against the sky. The reflection of the watch was seen by an eye-piece at the right focal distance. When it became perfectly distinct, the mirror had received its proper concavity. It was then placed in its box, lined with felt and pitch, so as to prevent any sudden change of temperature from greatly affecting its figure. It is singular that a nearly similar mode was devised for the bedding of specula by Lord Rosse in Ireland and by Sir John Herschel at the Cape of Good Hope. Sir John Herschel found that a speculum supported on three metalic points in the circumference made the image of every considerable star triangular, and that a packthread stretched down the back of the mirror for support distorted the images of the stars to a preposterous extent. He employed a great many thicknesses of blanket to prevent the effect of flexure in the wooden back of the case. To keep the elasticity of the fibre the blanket must be often shaken.

§ **72.** The speculum so fortunately completed by Lord Rosse was fixed or bedded on three iron plates, which gave it support, and then transferred to its appointed situation in the tube. This is three feet in diameter, and thirty feet long, and attached to an apparatus on the lawn, by which it can be brought to bear on any point of the sky from a short way above the horizon. The machinery for moving it round and raising and depressing it is simple and ingenious; and notwithstanding its size, it may be adjusted with the greatest ease. Two step-ladders form part of the apparatus, and by these we mount to a gallery, which can be raised or lowered to any required height. In order to procure an observation, the tube is first brought to bear on the star or other object, and the gallery being raised, we ascend to it by one of the ladders. On reaching the gallery, which is a small railed platform sufficient to hold several persons, we find ourselves close to the telescope, near its upper extremity; and here, on looking through a small eye-piece fixed to the tube, we at once recognise, in the obliquely-placed mirror within, the object of our observation. The tube is of wood, hooped with iron. The mouth of the tube remains permanently open. The telescope is lowered in wet weather, and the speculum is confined in a case, the cover of which is withdrawn by an exterior action when required. A vessel of quick-lime is also kept constantly in the case, for the purpose of absorbing the moisture and acid vapors, by which the speculum might be tarnished.

§ **73.** The eye-pieces used with this telescope range from 180 to 2,000 times the power of the naked eye. Unless the atmosphere be exceedingly clear, a powerful eye-piece will magnify the globules of watery vapor, and form a haze. Different densities from contending streams of cold and warm air have a similar effect; and if the atmosphere be exceedingly cold, as in a Russian winter, floating spicula of ice, invisible to the naked eye, are magnified so as to interrupt perfect observation.

Sir John Herschel found that the excessive heat and dryness of the sandy plains at the Cape often destroyed distinct vision, and that in a very singular manner. In some cases the images of the stars are violently dilated, and converted into nebulous belts or puffs of 10″ or 15″ or more in diameter. In others they form soft, quiet, round pellets of 3″ or 4″ in diameter, very unlike the spurious discs which they present when best defined, and rather resembling planetary nebulæ. Sometimes the structure as it were of these pellets is disclosed, and they are seen to arise from an infinitely rapid vibratory motion of the central point in all possible directions, while on a few occasions the appearances are exceedingly perplexing and singular. Some of the phenomena evidently have reference to the state of the air in the tube of the telescope; the tube of a reflector being necessarily open at the mouth, ascending and descending currents of hot and cold air, usually rotating spirally, are established, and are very prejudicial to distinct vision. The remedy is to dispense with a tube altogether, substituting for it a light, strong, inflexible, iron frame work. In refracting telescopes, where the air is completely inclosed, its circulation is not nearly so injurious.

§ **74.** The performances of Lord Rosse's telescope were found to be far beyond those of any previously constructed instrument. But Lord Rosse considered that something still grander could be achieved; and before the above telescope was well finished, he projected one of the extraordinary dimensions of six feet diameter in the speculum, with a tube of sixty feet long. The casting, grinding, polishing and mounting of this monster speculum were pretty nearly a repetition, on a larger scale, of what had been previously done. Its focal length is fifty-three feet; it weighs nearly four tons; and, as its diameter is six feet, it has an area four times greater than that of the three-feet speculum. When finished, the speculum was placed in a square box, which is attached to the lower end of the tube, and by means of a door can be entered at pleasure. This box adds six feet to the length of the tube, which, like its predecessor, is of wood, hooped with iron like a barrel, and so wide that a tall man could walk through it without stooping. This huge black funnel is suspended between high and strong walls. It swings with a clear space of twelve feet on each side; and so far it can be drawn aside, giving half an hour before and after meridian. By means of a windlass, and a most skilful adjustment of chains and counterpoising weights, it can also be brought to the zenith, or turned fairly round from south to north. Enormous as are its dimensions, and although weighing altogether twelve tons, it seems to be about as easily moved as the other telescope; and it is as much in the mechanical contrivances for effecting this purpose as in anything else that the peculiar merit of the structure consists.

CHAPTER IV.

NEBULAR AND SIDEREAL SYSTEMS.

The Milky way. Comparative dimensions of the Solar and Sidereal Systems. Distances of the Fixed Stars. Classification of Stars according to their apparent Magnitudes. Distribution of the Stars. Gauging of the Heavens by Herschel. True form of the Milky Way. Clusters and Nebulæ. Forms and distribution of Nebulæ. Vastness of the Universe. Effect of the finite velocity of Light.

§ **75.** In that portion of infinite space which is unveiled to the gaze of man lie clusters of countless suns, separated one from another by unimaginable intervals. Within one of these clusters, and probably nowise distinguished as to size or brightness from the other orbs, lies the sun around which our earth revolves. About this sun, at distances too small to be represented here, revolve the members of the solar system. With this bright company are we environed day and night.

Fig. 1, plate I, represents the outline of a section of the cluster to which it is supposed our sun belongs. The section makes an angle of 35 degrees with the earth's equator, crossing it in 124½° and 304½° of right ascension. A celestial globe adjusted to the latitude of 55° north, and having α Ceti near the meridian, has the plane of this section pointed out by the horizon. It cuts the milky way at right angles on one side in its two branches which cross the constellation of the Eagle, and on the opposite side in the southern part of the Unicorn towards the Canis Major. The circle in the figure includes all the suns or stars ever visible to the naked eye. On all sides of the earth, taken together, from four to eight thousand (for the number is differently estimated) may be seen. About two thousand may be seen by average eyes on an ordinary night in clear climates. In foggy island climates not more than nine hundred are visible at once. Inexperienced observers suppose the number much larger, partly because the sight is dazzled by their irregular distribution, and partly because as they diminish from stars of large size to those scarcely visible the imagination supposes others still smaller and invisible. The dot in the centre of the figure gives the position of our sun in the cluster. All stars beyond the circle, if they lie scattered in space are invisible to the unassisted eyes of the inhabitants of the earth. If they lie many in one direction, they present to the naked eye a milky, hazy appearance, and are called by the general name of nebulæ. This nebulous light belongs to distant groups of our cluster, and also to more remote clusters. Stars one hundred and eighty times the distance of Sirius are the most remote which appear even as nebulæ to the naked eye.

§ **76.** In order better to conceive of the dimensions of the solar system, and of our cluster, and the intervals which lie between the clusters, let us make ourselves familiar with known and moderate distances, and advance from these to those which almost baffle the imagination.

The sun is a globe 883,000 miles in diameter. A hollow sphere with a radius of three thousand millions of miles includes all the members of the solar system yet discovered. If the swiftest race horse had begun to traverse this sphere at full speed at the birth of Moses, thirty-four centuries ago, he would not yet have accomplished one quarter part of his journey.

§ **77.** Between the solar system and the stars lies a wide space traversed only by comets, and their appropriate field, if indeed they be not visitants from other spheres. The nearest fixed star, α Centauri, is twenty-one millions of millions of miles from the sun; 61 Cygni is fifty-six millions of millions distant from it. The distances of but few of the fixed stars have yet been ascertained. What we know of their distribution makes it probable that the stars of one cluster are on an average separated among themselves by distances as great as that between our sun and the nearest fixed stars.

The distances of those stars which have not yet been measured can only be inferred from their superior brightness. And here again for want of knowledge we must introduce another supposition. We must suppose that stars appear large merely in consequence of their proximity to us; and we must leave out of sight the differences which have lately been proved to exist in their actual size, or in the intrinsic brightness of their surfaces. As we know nothing of these particulars, and as they may vary in different stars in the ratio of many millions to one, we cannot be sure that we assign to any star its true distance. An arrangement of the stars in the order of their precise apparent brightness is much to be desired; but the variety of their color makes such an arrangement difficult. If they were catalogued according to the force of the whole impression made on the eye, we might obtain some knowledge about their intrinsic light-giving power, and might ascertain the extent of the changes which take place in the light of some of them.

§ **78.** At present the stars are loosely divided into classes according to their apparent size. All above a certain size are considered of the first magnitude; all less bright than these, and above a certain brightness, are of the second magnitude. Those decidedly inferior in brilliancy form the third class, and so on down to the sixth and seventh magnitudes, which comprise the smallest stars visible to the naked eye in the clearest night. Beyond these the telescope reveals new orders, and as higher space-penetrating powers are used, new orders are added. Of course the layer in which a star first appears does not give us its position in space, it may be a very large star and lie farther off, or it may be a small one and lie in the layer in which it first appears to us. All we expect to learn is the comparative brightness of the stars as seen from the solar system.

The division into magnitudes is arbitrary, nor is it easy to determine where one magnitude ends and another begins, since all those stars which are included in one magnitude are by no means of the same size.

The light of Sirius, the brightest of the fixed stars, is about 324 times that of an average star of the sixth magnitude. As might be expected, the number of stars of each magnitude increases rapidly as we pass from the first to the lower magnitudes. There are from fifteen to twenty stars in the first class, and, unfortunately for us in the northern latitudes, the largest and most brilliant of these are not visible in our heavens. Of the second magnitudes there are fifty or sixty stars; of the third, 200; and in the first seven classes taken together, there are upwards of 2,000 in the northern hemisphere; in the milky way about 18,000,000; and in all the nebulæ, about 100,000,000 distinct stars are within reach of telescopic vision.

§ **79.** The three or four brighter classes are distributed with tolerable equality throughout the heavens, but the smaller ones visible to the naked eye increase rapidly in number as we approach the borders of the milky way. The telescopic stars are crowded beyond imagination along that circle and the branch which it sends off, so that its whole light is composed of stars, whose average magnitude is not above the eleventh or twelfth. It was computed by Herschel that in one hour 50,000 passed through the field of his telescope in a zone 2° in breadth. This compression was partly owing to the vast numbers brought within his line of vision in depth, and partly to the real crowding of the stars in the milky way.

This unequal distribution of stars enables us to learn approximately the form of our cluster. Sir William Herschel determined it on the following principle. If you were in a crowd or immense building filled with people, you would judge your distance from the edges of the crowd or from the walls of the hall from the number of people seen in each direction. If you were in a wood and wished to determine its outline without leaving its interior, you might form a rude approximation to the true outline by taking your position in one spot and drawing imaginary lines in every direction to the edge of the wood. If one hundred trees were visible in one direction, you might assume for the line running thither a certain length, and proportion all your other lines to this, making them longer or shorter as more or fewer trees were visible in each direction. A bounding line passing through the termination of each of these lines would be not far from the true outline. A body which has extension in three directions, as our cluster, may be treated in the same manner as the wood. Herschel used the telescope as a sounding line, and inferred from its discoveries the hollows, protuberances, and in fact the shape of a great portion of our cluster. He made 700 observations to fix its form and dimensions.

In order to determine the comparative mean richness in stars of any two regions of the firmament, Herschel made use of a telescope which magnified 187 times, and whose field embraced a circle of 15′ diameter. This field included each time about the eight hundred and thirty thousandth part of the entire heavens. Towards the middle of the first of these regions he counted successively the number of stars included in ten fields contiguous or at least very near each other. He added these numbers and divided the sum by 10. The quotient was the mean richness of the region explored. The same operation, the same numerical calculation, gave him the mean richness of the second region. When the latter result was double, triple, or tenfold the former, he inferred that a stratum of it contained twice, three times, or tenfold as many stars as a stratum of the former of equal depth, and consequently that the stars extended twice, three times, or ten fold as far in the latter as in the former direction. In some portions of the sky at least four successive fields were required to meet with three stars. Elsewhere these circular areas of 15′ diameter contained 300, 500, 588 stars.

To sound the whole heavens in this way would require more than a million of observations. Herschel made 700 soundings, omitting however the circumpolar regions. He did not therefore learn with certainty the whole form of our cluster.

Suppose the whole sphere, as far as the eye can reach, to be represented by a common two-feet celestial globe. Herschel sounded what would be in proportion to this globe one hundred and fifty feet beyond its centre, and made a chart of it in section extending to a proportional distance from our sun. On the same scale, a chart of the discoveries possible to Lord Rosse's telescope, would have for the radius of the sphere from which it was taken one thousand feet. Herschel's observations, though taken from the earth as a centre, apply equally to the sun as a centre, for the ninety-five millions of miles which separate the sun and the earth make no more difference in the position of those bodies than the distance between two observers who side by side watch the setting sun makes in the position of that luminary.

§ **81.** Studying thus the heavens, Herschel found on most sides only a small number of large stars ; but toward the bright belt called the milky-way, are myriads of stars, so distant as only to be visible by a faint white light.

The increase of numbers in approaching the milky-way is imperceptible among stars of a higher magnitude than the eighth, and except on the verge of the milky-way itself, stars of the eighth magnitude can hardly be said to participate in the general law of increase. For the ninth and tenth, the increase, though unequivocally indicated over a zone, extending at least 30° each side of the milky-way, is by no means striking. It is with the eleventh magnitude that it first becomes conspicuous, though still of small amount when compared with that which prevails among the mass of stars of magnitudes inferior to the eleventh, which constitute sixteen-seventeenths of the stars within 30° on each side of an imaginary circle running through the middle of the milky-way. Two conclusions follow from this; first, that the larger stars are really nearer to us than the small ones; secondly, that our system is plunged in the sidereal stratum constituting the galaxy, to a depth equal to about that distance, which corresponds to the light of a star of the ninth or tenth magnitude, and certainly does not exceed that corresponding to the eleventh.

Applying to this cluster the principle by which the wood was mapped, it must be by no means globular, but rather like a slice through the centre of a globe, extending farthest in the direction of the milky-way. Its major axis is estimated at seven or eight hundred, and its minor axis at a hundred and fifty times, the distance of Sirius. It is however by no means certain that Herschel sounded to the limits of our cluster.

§ **82.** Our cluster has been compared in form to a grindstone, a little bulging instead of plane on the sides, and having its rim split through about one third of its circumference, but not nearly to the centre, the parts dividing at an angle of about 30°. A section of this at right angles to the plane of the circumference would have nearly the form of the letter Y. Our sun is just below the cleft or the joining of the Y, near the vertex of the angle, nearer to Sirius than to the Eagle, nearer to the Southern Cross than to Cassiopeia, and almost in the middle of the starry stratum in the direction of its thickness.

Suppose the grindstone to be very porous. Let its minute atoms represent stars, the pores between being the interstellar spaces. An observer within such a cluster would have a scene resembling our own celestial vault. Toward the sides comparatively few stars would be seen; toward the circumference a succession of remote stars would form a zone lying like our milky-way. Thus on our blank sides only forty stars in succession are seen, in the direction of the milky-way nine hundred may be counted, though this is not in all parts equally profound.

Between it and the central portion of our cluster are breaks and vacuities, which detach it and make it appear more like a separate ring of stars. The ring in many parts consists of separate groups of stars, mostly spherical in form, the groups sometimes lying close together, sometimes having spaces between. In some parts there is an average of 3,138 stars in a square degree, and in the denser part 5,093 in the same area. It varies in breadth from 5° to 10° and even 16°. Its telescopic breadth is 6° or 7° greater than that assigned by the naked eye, the stars becoming much less numerous near its edge. Even the apparent breadth of 5°, in its most distant parts of 900 units of distances of Sirius gives a real linear breadth of 78 units of distances of Sirius. Therefore the milky way greatly exceeds even in breadth the reach of the naked eye. It twice separates and unites. Its cleft part extends from below the ecliptic in the southern hemisphere to very near the pole of the ecliptic in the northern hemisphere.

§ **83.** Its two branches between Serpentarius and Antinous separate more than 22° of the sphere. It is inclined to the earth's orbit about 60°, and its edge is not more than 22° distant from the poles of the ecliptic. In northern latitudes it is most conspicuous from July to November. It is visible by night at all seasons, in all positions of the globe. As seen by us projected on the celestial sphere it differs but a few degrees from a great circle, we being almost in its plane. If we were more on one side of its plane it would appear like a small circle, and would divide the heavens into two quite unequal parts. This gives us our position as to one dimension of our cluster.

We learn from other appearances that our north pole is farther from the milky way than our south pole; that is, that we are nearer that part of the ring which approaches our south pole than that which approaches our north pole. Not only does the telescope give more soundings toward that part of the ring which encircles our northern hemisphere, but the whole light in the northern hemisphere is fainter, and evidently comes from exceedingly remote stars. In the southern hemisphere not only is the belt a blaze of light, but many more stars of the first magnitude are discernible, proving our greater proximity to them. From Sirius to Antinous it is perfectly illuminated with stars, many of which are visible to the naked eye, while toward the north the light gradually becomes hazy and without a trace of stars. The very great size of the southern stars makes it probable that we lie not very far inside the detached ring.

§ **84.** As from the contemplation of the solar system we stretched our imagination to embrace the fixed stars, we must now pass on and conceive these myriads of stars forming a cluster, isolated in space, and separated from other clusters by intervals in most cases far greater than the distance between our sun and the most remote portion of the milky way. These clusters seen through the telescope vary in size, brightness, number, and color of stars, and present a variety of fanciful shapes. The elliptic form however prevails with an increased brightness in the centre greater than would arise merely from the depth of the cluster, and which must be attributed to the greater proximity of the central stars owing to their mutual attraction. The stars which compose the clusters also show the influence of gravity by their disposition to break into groups.

The nearer clusters offer to the eye only a faint diffused light. Viewed through a telescope of moderate power, they resemble a handful of fine sparkling sand, or, as it is called, star dust. A higher power brings distinct stars into view at small intervals, and with a faint light. The stars near the edges may be distinctly seen, while those near the centre unite their rays and form a brilliant light. A higher power enables us to see individual stars with great distinctness. Many clusters contain ten or twenty thousand stars wedged together in a space whose area is not more than one tenth part of that of the moon, so that the centre where the stars are seen one behind the other is a blaze of light.

Herschel applied to these clusters the principle by which he had ascertained the distance of the stars of one cluster.

He thus ascertained that the distances of forty-seven resolvable clusters were at least 900 times that of Sirius from the sun. A cluster first resolved by Lord Rosse's mirror must lie six thousand times as remote from us as the nearest fixed stars.

§ **85.** Before this immense power was turned upon the nebulæ it was supposed that besides those clusters of stars, which from their distance take a nebulous appearance, there was floating in the sphere a very thin nebulous substance, perhaps the material from which stars were condensed. The immense size of some nebulæ seemed to make it improbable that they could consist of clusters of stars. Lord Rosse's telescope has, however, resolved some of the largest and faintest of these nebulæ, and thus made it probable that all of them may yield to yet more powerful instruments. The distinction, therefore, between nebulæ and clusters does not properly exist in nature, but probably arises from the low power of the instruments used. When the stars which compose a cluster are so small, or so close, or both, as not to be separable by the telescope, the cluster may offer every variety of illumination from a mere vaporous patch of light to a brilliant surface of mottled or even of sensibly uniform illumination. In this case, if the nebula is very distant, it may present a uniform disc like a planet, more or less well defined and uniform, and will be called a planetary nebula. These objects are seen in both hemispheres, but less so in the southern than in the northern. If, however, the nebula be not only globular, but compressed towards the centre, when exceedingly remote it will have the appearance of a single star surrounded by nebulous matter. Sometimes nebulous matter appears appended to a nucleus like that of a comet. In this case the cluster is irregular and the condensation not central. For the present, however, the division of nebulæ into resolvable (clusters) and irresolvable, facilitates their description.

§ **86.** Nebulæ are also divided as to their form into regular and irregular; the regular form being elliptic, with a diminution in the ellipticity of their strata from without inwards, so as to approach a spherical nucleus, however elongated may be their outline. The regular nebulæ appear globular to all but the most powerful instruments, because the outer layers are usually faint. From this cause it is generally asserted that the globular form prevails. Annular nebulæ are considered regular, for they are probably hollow spherical shells of stars. Planetary nebulæ and globular clusters also come under this head.

Irregular nebulæ are the most extensive objects in the heavens. Their forms are most capricious, imitating a dumb-bell, a fan, and in one instance a human head and breast. Their true form is not however seen by us, and it may be that many nebulæ lying nearly in our visual line are by our eye blended into one. Probably in some cases they really consist of systems of systems. Indeed double nebulæ, or those lying very near one another, occur so frequently that their proximity cannot be attributed to chance. The nebulous system appears to be distant from the sidereal, though involving it, and perhaps to our eyes intermixed with it. The limits of our sidereal system have not in all directions been defined. The distribution of the nebulæ is not like that of the milky way in a zone or band. One third of the nebulous contents of the heavens are congregated in a broad irregular patch occupying about one eighth of the whole surface of the sphere, almost entirely situated in the northern hemisphere, and occupying the head and shoulders of the Virgin, and the surrounding constellations. Within this region are several local centres of accumulation where the nebulæ are exceedingly crowded. The lesser nebulous region of the northern hemisphere lies in Pisces, and is much less concentrated.

§ **87.** In the southern hemisphere a more uniform distribution prevails. With the exception of the two Nubeculæ, which are full of nebulæ, and the greater of which is even richer than the denser portion of the northern group, this hemisphere contains alternating patches of nebulæ and vacuities of greater or less extent. In one of the vacuities in which comparatively few nebulæ occur the south pole is situated, having one nebula, however, within half a degree of it, as the north pole also has one within five or six minutes. This barren region extends nearly 15° on all sides of the pole, and immediately on its borders occurs the smaller Nubecula. One of the most remarkable features in the southern nebulous system is the extraordinary display of fine resolvable and globular clusters which occur in the region occupied by Corona Australis, the body and head of Sagittarius, the tail of Scorpio, with part of Telescopium and Ara. Here in a circular space of 18° in radius are collected no less than thirty of these beautiful objects. Are we to suppose them to be a bunch of general nebulous systems nearer to us than the rest? Or is it merely that on this side we approach nearer the milky way? It cannot be doubted that some of these objects form a part of the milky way.

§ **88.** The bright fleecy spots, long known to mariners as the Magellanic clouds, are composed of large patches of unresolvable nebulæ, and of nebulosity in every stage of resolution, up to perfectly resolved stars like the milky way, as also of regular and irregular nebulæ, properly so called, of globular clusters in every stage of resolvability, and of clustering groups sufficiently insulated and condensed to come under the denomination of clusters of stars. The Nubecula Minor contains within an area not much exceeding ten square degrees, forty-three nebulæ and clusters; the Nubecula Major, within an area of about forty-two square degrees, contains 278, without reckoning fifty or sixty outlines, making an average of about six and a half to the square degree, which very far exceeds any thing that is to be met with in any other regions of the heavens. This intermixture of stars and unresolved nebulosity makes it probable that the nubiculæ are systems which resemble none in our hemisphere. Sir John Herschel has ascertained the places of 919 stars, nebulæ and clusters in the Nubecula Major, and of 244 in the Nubecula Minor, as an approximation toward a catalogue of the objects they contain. He has also fixed the places of 4,015 nebulæ or clusters, of which the southern hemisphere contains the larger portion. Each hemisphere contains about as many as the eye sees stars on an average night.

Numerous and vast as these clusters are, the distances

which separate them are yet more astounding. Imagine clusters of suns, each sun lying so far distant from the other that the eye can pass over only six such intervals in one direction, while the cluster contains from end to end hundreds of such suns lying at such intervals. Then imagine these clusters lying so widely separated from one another that they are but as handfuls of dust in space. How wide must be that universe of which man cannot comprehend a corner!

§ **89.** It may assist us to realize their vast distances if we consider how long light travelling 192,000 miles a second would be in travelling from them to us. Light is one and one quarter seconds passing from the moon to the earth; eight minutes from the sun; three to twelve years from the nearest fixed stars; 140 years from the most distant stars visible to the naked eye; thousands of years traversing our cluster in its longest direction, from Aquila to Monoceros; and millions of years coming from distant clusters, a period long enough to allow important changes in the cluster from which it emanates. Thus the moon may have been dispersed into atoms for more than a second, and the sun for eight minutes, and we should still see them perfect and entire. The star α Centauri may have changed its color three years ago, and we should still see it of its former hue. The bright star Vega, must have been placed in the heavens nine years before its rays struggled to our little world; and more distant stars may have been shining for centuries yet not so long that their light has reached the earth. The light which now meets our eyes may come from stars long since quenched in darkness. A human being may be born, pass through the seven stages of life and die, while light from the smallest stars visible to the naked eye is reaching us. Nay, our whole historic period is about the length of time which light occupies coming from the nearer or cleft edge of our cluster to the earth. Thus the astronomer who records the aspect and variations of a distant nebulæ gives its history millions of years since. If the solar system and the fixed stars were called into existence at the same moment, from the earth no other body would at first have been seen.

The moon would have appeared in a second and a quarter, the sun in eight minutes, the stars would have peeped out one by one in the course of years ; there would have been no field for the telescope under a century. An exact chronicle of their times of appearing would have been a perfect measure of their respective distances from us.

CHAPTER V.

INTERIOR OF OUR SIDERAL SYSTEM.

Absolute and Relative Motion. Motions of the Fixed Stars. Proper Motion of 61 Cygni. and of Arcturus. Motion of the Solar System. Investigations of Herschel, Struve, and Argelander. The Central Sun. Double and Multiple stars. New Stars. Variable Stars. Color of Stars.

§ **90.** Having obtained a general idea of the form of our cluster, and of its distance from other clusters, we will now study its interior. We will inquire whether the bodies which compose it move among themselves, and whether they undergo any changes of constitution ?

Of absolute motion we can know nothing. Relative motion is all it concerns us to know. But we want some fixed point to which we may refer motions ; and the heavens afford us no such fixed point. The cluster of which our sun is a unit may, for all we know, be rushing on with unimaginable speed ; but its suns retaining their relative position would still appear to be at rest. Our sun may be changing his place among the stars, but only centuries of observation can make his motion evident. The earth is certainly revolving round the sun and rotating on its axis. These two latter motions we perceive by reference to the stars, which do not partake of them, but how are we to ascertain any motion we have in common with the stars.

We know not whether there be in the universe one star deserving to be called absolutely fixed, and we must be contented to refer our motion to the bodies nearest us.

§ **91.** By observing and reasoning we shall be able to ascertain whether the motions discovered are entirely our own, entirely belonging to the stars, or compounded of both. We have instances of all these kinds of apparent motion in the heavens. The rising and setting of the sun and its yearly motion are entirely apparent and owing to the earth's motion. The moon's motion is made up partly of our real motion, partly of her revolution; and besides this the moon and earth share a third motion round the sun. We do not see in the solar system a motion which is not influenced, either accelerated, delayed, or changed in direction by the earth's own motion. When we inquire into the motion of the fixed stars, the earth's proper motion no longer embarrasses us; it is too small to be of the least account; but we must allow for possible motion of the sun and solar system together. We must inquire whether this motion belongs to them, or to us, or to both. We must observe whether the apparent motion is common to all the stars, whether it is in such a direction that the motion of our sun would account for it, or whether absolute and parallactic motions unite in producing the apparent motions of the stars.

§ **92.** The fact that the fixed stars change place among themselves has long been suspected. Maps and observations made at intervals of fifty years differ. So that it may be affirmed with certainty that a map of the heavens, correct this year, will, after a few years, and still more after a few centuries, be found faulty. The immense distance of the stars causes their motions to appear slight, or not to appear at all to the naked eye. Millions of miles of their path subtend to our eyes not even an angle of one second only. The annual motion of sixty-one Cygni is more than a thousand millions of miles, yet we call it a fixed star. To the eye of a common observer the heavens present the same features they did thousands of years ago. But by comparison of catalogues many minute changes may be detected, and modern instruments can measure their changes from year to year.

As might be expected the larger and nearer stars show the most motion. This would be the case whether the motion were theirs or ours. The bright star Arcturus is moving towards the south-west with a velocity of two seconds and a quarter of arc every year. Sirius, Aldebaran, Castor, and others, are likewise rapidly moving.

§ **93.** Since so many of the neighboring stars move, it is probable that our sun, similar in its nature and subject to the same laws, also moves, and that the motions we see in the stars are the differences between their motion and ours. If the stars are at rest our sun must move. Either hypothesis gives the sun proper motion. It remains then to ascertain toward what point the sun is moving, and this may be done by applying a very simple principle. As we pass among columns, or the trees of a forest, those we approach separate from one another. while those we leave behind seem to close together. The greatest apparent motion is in those stars we are overtaking and leaving behind, those which are at right angles to the sun's motion. Herschel judged that the sun was moving towards a point in the constellation Hercules, because the distances between the stars in that region are becoming greater. This conjecture has lately been confirmed by the investigations of Argelander and of Otto Struve.

§ **94.** Argelander compared the positions of 560 stars in all accessible regions of the heavens with those laid down in the catalogues of preceding astronomers. Of these 560 stars, 170 moved so slowly as to yield no reliable results in the short time between the first and last observations. Of the remaining 390, the slowest motion gave a yearly change of place amounting to one tenth of one second of an arc. These stars were arranged in classes according to the rapidity of their motions. The first class included all whose yearly motion equalled or exceeded one second of an arc, and contained 21 stars. The second class included those whose yearly motion exceeded one half, and fell short of a whole second. This class contained 50 stars. The third class comprehended all whose motion was between one half and one tenth of a second, and contained 317 stars.

§ **95.** By a rough examination it became evident that the sun belongs to the class of rapidly moving stars, and that the solar motion was directed toward some point in the constellation Hercules, where indeed the elder Herschel had placed it.

The determination of the exact point was now a matter of trial. Argelander examined the motions of the stars of the first class. He selected a point in the constellation Hercules toward which he supposed the sun moving, and computed the direction in which these stars would seem to move if their apparent motion were caused by the sun's real motion toward the selected point. The computed direction was compared with the observed direction, and if they coincided the computed one would be considered the true one. If there were differences, a point must be selected which would reduce these differences to the least possible amount. The three classes of stars were investigated in this manner, and the point in Hercules, toward which the sun was moving in the epoch 1840, was ascertained.

§ **96.** An attempt has even been made to calculate the velocity with which the sun moves in its orbit. The star 61 Cygni is known to be 660,000×95,000,000 miles from the sun, and its apparent motion through space has been accurately determined. Its distance from the point of observation being known, that apparent angular motion can be converted into an apparent motion of so many miles. Now if this displacement of 61 Cygni is owing to the translation of the sun, the velocity with which we are darting through space may easily be ascertained.

If so we might move the distance of 61 Cygni in 40,000 years. In five hundred thousand years we might reach the extremest verge at which the eye can descry a single star. In two hundred and fifty millions of years we should reach the remotest distance to which Lord Rosse's telescope can pierce.

§ **97.** But this hypothesis of the sun's motion by no means accounts for all the motions observed among the stars. The next question is, in obedience to what law the sun and stars perform their proper motions. Then again we want long continued observation of the stars, and comparison of catalogues made at great intervals, in order to learn the precise direction and quantity of the motion. The law which guides the binary systems, and which appears to influence even the form of the clusters is the law of gravity; we may therefore suppose it to regulate the stars of each cluster.

Reasoning from analogy with our solar system, we may suppose the cluster to contain a great central luminary around which all are tending. Reasoning from analogy with the multiple stars, we may suppose all to move round some central point.

§ **98.** Mr. Maedler, of Dorpat, thinks he has discovered this central point. The following is an outline of his reasoning. Since gravitation extends throughout our system, there must be a centre of gravity to it. What fills this centre, or whether it is filled at all, becomes a matter of special research. Admitting the existence of one grand central body, visible or invisible, predominant over others, we should expect to find the most rapid motions in its immediate vicinity. That many of these revolving bodies and even the central body itself may be invisible, will not be denied for a moment; but as we are able to look out upon the starry heavens in every direction, we ought to find a point in which the most rapid motions are concentrated, and from this point outwards the quantity of motion should continually decrease. That an apparently more rapid motion than that found near the centre cannot be predicated of any stars except those very near our sun is manifest.

§ **99.** If however we adopt the hypothesis of the centre of gravity merely, the proper motions of the stars in the neighborhood of it will be feeble. In this direction rapid proper motions can exist only in those stars in which their great proximity to our sun may seemingly increase their velocity. The stars beyond the central point will exhibit only slow motions, since their great distance from our sun will counteract their actually increased velocity.

In opposition to the great centre the motions would be swifter than at our sun. But since only the difference of their proper motions and the sun's will in most cases be known to us, and this only at very considerable actual distances, we must only expect feeble motions in opposition.

More rapid proper motions are to be sought for only at great distances from both of these points. How much is to be expected in each case will depend upon two circumstances; the distance of our sun from the central point, and the direction of the star's path.

§ **100.** Maedler considers the course of the milky-way, according to the determination of other distinguished astronomers, as pointing out the plane, which is to be taken as the ground plane of the starry stratum, and he fixes the central point in this plane. This course describes nearly a great circle, in the heavens, yet with a deviation not to be neglected, since the milky-way under the meridian of the vernal equinox, at a mean, is 36½° from the north pole, while under the meridian of the autumnal equinox it is only 26½° from the south pole. This deviation appears less clearly in the points of intersection of the milky-way and the equator, as a consequence of the double division (of the milky-way) on one side. It would moreover seem that of the two parts into which the milky-way divides the celestial sphere, that one is *least* in which the vernal equinox falls, and is farthest from us, our sun being on the outside of this plane towards the side of the larger part, and we are to seek for the central point toward the smaller portion.

Both the Herschels have demonstrated that we are nearer the southern half of the milky-way than the northern, as is seen by a comparison of the depth of the stars in the two directions. Through this circumstance we are in a condition to confine our researches for the central point to narrower limits, since we must seek it in the northern part of this smaller portion and between the milky-way and the equator.

We may, however, add that if our sun is not very near the central point, (as is far from probable according to Herschel's researches on the different depths of stars in the milky-way,) on the side of the central point a greater crowd of stars is to be expected, although they may not be as dense as those which the region of the milky way presents. All regions not thickly strewn with stars we may reject. None, however rich and promising, can be admitted as the true centre unless the proper motions of the stars observed and registered point toward it.

§ **101.** From a close examination of the catalogues, Maedler has reached the conviction that in the Pleiades the central group of the entire starry heavens included within the milky-way is found. He even selects a bright star of this constellation, Alcyone, as the individual star more likely than any other to be the true central sun. Among the compressed groups of the heavens no one is found which approaches the Pleiades in brilliancy and richness. It is also located in a region very rich in stars, and at a point which fulfils very completely the required conditions.

Maedler makes however the reservation, that, as in our system, notwithstanding the great preponderating mass of the central body, the centre of gravity falls without it so often as Jupiter and Saturn are separated by less than one quadrant from each other; in like manner the centre of gravity of the starry heavens, in the changes among the constellations in the lapse of thousands of years, may fall without Alcyone, or even pass over to another star.

§ **102.** Assuming Bessel's parallax of the star 61 Cygni to be correctly determined, Maedler proceeds to form a first approximate estimate of the distance of this central body from the solar system. And he arrives at the conclusion that Alcyone is about 34,000,000 times as far removed from us or from our sun as the latter luminary is from us.

The same approximate determination of distance leads to the result that the light of the central sun, if there be one, occupies more than five centuries in travelling thence to us. The enormous orbit which our own sun is thus inferred to be describing about the distant centre, not indeed under its influence alone, but by the combined attraction of all the stars which are nearer to it than we are, and which are estimated to amount to more than 100,000,000 times the mass of our own solar system, is supposed to require upwards of 18,000,000 years for its complete description, at the rate of eight miles for every second of time. These calculations are however far from conclusive. They do not account for all the proper motions of the stars, many of which are in opposite directions, showing rather the gravitation round several centres than one universal centre.

§ **103.** The motions we have described are not the only ones discoverable in the stars. Some of the stars revolve in pairs, or in groups of four or five, round a common centre of gravity. These are called binary and multiple stars. From the change of brightness in stars apparently single, motion too small to be otherwise detected is inferred.

Some of these stars entirely disappear, and some vary in brightness periodically, while others are irregular in their changes. The star Algol passes through a change which occupies two days and twenty-two hours. The second star in the Lyre goes through its variation in six days and nine hours. A star in the Swan, from being visible to the naked eye becomes invisible and resumes its former brightness in the course of eighteen years. Of the stars which are irregular in their changes, some disappear suddenly, others undulate in their variations, increase in brightness, diminish, and again increase.

§ **104.** Catalogues have not heretofore been sufficiently accurate to determine how many stars have reappeared in the same places, and how many are to be regarded as new stars. It is certain that many formerly recorded in catalogues have disappeared, and that many new comers have become permanent dwellers in the heavens, while others have shone a while with the utmost brilliancy and then disappeared. Observations during the last fifty years, show ten or fifteen stars which have either entirely disappeared from their former places, or so completely changed their magnitudes as no longer to be registered. Twenty have changed in brilliancy, either greatly increased or diminished. Several have become visible of a size and position too striking to have been overlooked if before so bright. One of the fifth magnitude, and consequently visible to the naked eye, has made its appearance in the present year, 1848.

Sir John Herschel gives the following account of the changes which the star η Argus has undergone within a few years. In a catalogue of the stars made in 1677, it is marked as of the fourth magnitude; in subsequent catalogues, it is put down as of the second. When first observed by Sir John Herschel, in 1834, it appeared a very large star of the second, or a very small star of the first magnitude, and thus remained without any apparent increase or change up to nearly the end of 1837. On the 16th of December, 1837, Sir John Herschel was struck by the appearance of a very bright star in a part of the heavens with which he was perfectly familiar. On examination it proved to be η Argus, shining however with a light nearly tripled. From this time until the second of January, 1838, it continued to increase. After this it decreased rapidly. Again, in 1843, it increased, and in 1845, when last particularly observed, was again on the decline.

§ **105.** There are now seven or eight authentic records of the sudden appearance and subsequent extinction of new and brilliant fixed stars. They have once or twice appeared so suddenly as to strike the eye even of the multitude. One of the most remarkable instances occurred to Tycho Brahe. On Nov. 11th, 1572, as he was walking through the fields, he was astonished to observe a new star in the constellation Cassiopeia, beaming with a radiance quite unwonted in that part of the heavens. Suspecting some disease or delusion about his eyes, he went up to a group of peasants to ascertain if they saw it, and found them gazing at it with as much astonishment as himself. He went to his instruments, and fixed its place, from which it never afterwards appeared to deviate. For some time it increased in brightness—greatly surpassed Sirius in lustre, and even Jupiter; it was seen by good eyes even in the day time, a thing which happens only to Venus under most favorable circumstances, and at night it pierced through clouds which obscured the rest of the stars. After reaching its greatest brightness, it again diminished, passed through all degrees of visible magnitude, and finally disappeared. Some years after, a phenomenon, equally imposing took place in another part of the heavens, manifesting precisely the same succession of appearances. We are quite baffled to account for these astonishing displays. If the bodies in question are moving in orbits, how singular, that no change of position was observable, and how tremendous the velocity which could sweep these stars in so brief an interval from a region comparatively near to us to the invisible depths of heaven. From a comparison of records, there is some ground for supposing that the star seen by Tycho is not a stranger, but one which appeared before, passing through its mighty phases in about 300 years. If this be so, it ought to appear in about twenty or thirty years.

§ **106.** Many of these changes may be accounted for by supposing stars to move in orbits more or less elongated. Some of the orbits appear nearly circular; others resemble those of comets; some occupy a few days; others cannot be passed through in less than centuries. If a star were receding from us in a straight line we could detect its motion only by the diminution of its brightness, and if it recedes sufficiently far it may become invisible. If it moves in an orbit, it will appear again near its former place, and perhaps be regarded as a new star. The lost Pleiad may thus have receded to a distant part of its orbit, and may hereafter reappear. When the period is extremely short, and the star disappears, it has been suggested that, reversing the law of our system, it circles round a non-luminous body. When the period of a star occupies some years, we have no difficulty in attributing its dimness or disappearance to its greater remoteness. But when the period of a star's variations occupies only three days, it seems probable that some other cause of the variation in light exists. Perhaps further observations will show that for these various phenomena a variety of causes evists. Two other causes have been suggested besides those already mentioned. One is the existence of cosmical clouds, which may for years lie between a star and us, and dim its light; the other is the occurrence of changes at the surface of the star. Both of these explanations have this recommendation, that they account for the changes we see in the color of the stars as well as those which take place in their brilliancy.

§ **107.** The most careless observer must be struck with the variety in the hues of the larger stars, and the telescope reveals similar differences in all. Some are of a deep blue, red, or yellow, and not the least beautiful are of a clear sparkling white. In the southern hemisphere there are two planetary nebulæ which have of themselves one a pale but decided, and the other a more striking and intense blue color. Looking back to a great distance of time we find the color of some of the stars has changed. In the time of Plotemy, Sirius was classed with the five other red stars, Arcturus, Aldebaran, Pollux, Antares, and α Orionis. It is now of a brilliant white. Clouds of some partially opaque substance may have given Sirius its red color, as they may also cause the variations in lustre of those stars which do not change periodically.

§ **108.** The phenomena both of motion and color may be best studied in the double stars. To these curiosity has directed the telescope ever since its first invention. The places of many thousands have been ascertained. The distribution of the stars was always a question of great interest. Astronomers sought to know whether they were related, whether they revolved round one another, or round one great centre. It was early observed that many stars were nearer to one another than they would be if they had been accidentally scattered in the heavens, thus making it probable that some law caused their proximity. The telescope showed that some of those stars whose brightness particularly attracts us, consist of two stars very near one another and which appear but one to the naked eye. In some cases the proximity is only apparent, they lie one behind another, nearly in the same visual line to us, but separated by millions of miles. But double stars occur in too great numbers for us to suppose a mere coincidence of direction, and the great law of gravity enables us to account for their proximity. In most cases two stars, called binary or double, revolve round their common centre of gravity. Sometimes what to the eye appears one consists of three or more such stars, in fact is a well-balanced republic, moving harmoniously round the common centre.

§ **109.** Computations have been made of the positions of the double stars, allowing that they move in an orbit governed by gravity. The theoretical and observed positions are then compared. Not more than one or two have completed their revolutions since they were first observed, and there is no sufficient evidence that the same orbit has been retraced in successive revolutions. The periods assigned to them are from forty to twelve thousand years.

In studying the stars which move in orbits, the embarrassment is slight compared with those which either move with us onward or backward, or take their apparent motion from ours. A sun moving round a dark centre would describe an ellipse. If this ellipse were at right-angles to the radius of our visual sphere, we should see it in its true proportions. If it were inclined differently, it would be foreshortened, and the ellipse would change its form. If it were in the same plane with the radius, the star would appear to vibrate back and forth in a straight line. From any of these appearances the actual form of the ellipse could be ascertained.

When both stars are visible, as in the binary systems, it is a little more difficult to conceive of the motions. Each star moves round the common centre of gravity, and this common centre is the focus of each ellipse. The two stars being usually of unequal size, describe unequal ellipses, but their mutual attraction makes these ellipses similar, and causes the bodies to be in similar points of each ellipse at the same moment, and consequently to lie opposite to one another in a straight line drawn through the centre of gravity. Motion has also been detected in the systems of three, four, five, and even six stars.

§ **110.** Binary stars are among the most interesting objects in the heavens. The two stars are seldom of the same color, and rarely of colors at all similar. If one is red, the other is usually green; if one is white, the companion is dusky. They are also usually unequal in size. The large star is usually orange-red or yellow, and the small star blue, purple or green. Imagine the state of the planets, if any there be, attached to these suns! Think of being exposed to the alternations of blue and yellow days, and to every variation of color arising from the presence of one sun, of the other, and of the two together!

These colors were at first supposed to be an optical illusion. A ray of white light contains all the colors of the spectrum. When the eye is fatigued by gazing at a color, a ray of white light falling on it finds it insensible to that color. It excites, therefore, the perception of that color only which the eye is capable of perceiving, that is the color most unlike the other, its complementary color. The apparent color of some of the stars may be owing to this fact, but in many cases the color of each star remains unchanged when viewed alone. Whatever be this mysterious power by which stars thus connected seem to divide the spectrum, there can be no doubt of its existence.

In α Centauri, the most superb double star in the heavens, both stars are of a ruddy or orange color, though the smaller has a more sombre hue than the larger. The principal star is of the first magnitude, the companion is between the first and second. When the light of this double star is exceedingly magnified, it is difficult to assign its exact size to the second star; it is apt to appear larger than reality. If the light of both be weakened by reflection, the difference between them is more evident. The diameter of the relative orbit of these stars about each other cannot be so small as that of the orbit of Saturn about the sun, and exceeds, in all probability, that of the orbit of Uranus.

CHAPTER VI.

THE SOLAR SYSTEM.

The Primary and Secondary Planets. Law of distances from the Sun. Revolutions of the Planets in their Orbits, and their revolutions on their Axes. Elements of their Orbits. Amount of Solar heat received by each Planet. Orbits described by the Secondary Planets. Mass and Densities of the Sun and Planets. Eclipses. General views of the Solar System.

§ **111.** Let us return from these distant orbits to the only fixed star with whose immediate neighborhood we are acquainted. The other stars may or may not be accompanied with silent and invisible attendants, but we know that our sun is surrounded by a train whose minute size and dim reflected light must conceal them from distant observers whose organ of vision or whose telescopes are not greatly superior to our own.

Eighteen primary and twenty secondary planets compose the known solar system. It contains no other bodies of any considerable mass or they would disclose themselves by disturbing the motions of the planets. There are however two sets of bodies, meteors and comets, which have some claim to be considered members of the system ; and which will be described hereafter.

§ **112.** We are now more particularly interested in the primary planets, of which the earth is one, and the secondary planets or moons which revolve round their primaries while they accompany them round the sun. Seven of the primary and two of the secondary planets have been discovered within the last three years, and better instruments and more extended observations may hereafter reveal others.

Beginning with the planets nearest the sun, they are Mercury, Venus, the Earth, Mars, Juno, Ceres, Pallas, Vesta, Astraea, Iris, Hebe, Flora, Metis, and Hygeia, called from their small size, the Asteroids ; Jupiter, Saturn, Uranus, called also Herschel, and Neptune.

The Earth is accompanied by one moon; Jupiter by four; Saturn by eight, and by two or more rings; Uranus by six and perhaps by more; Neptune by one, and probably more. The planets near the sun are so involved in its rays, and those extremely remote are so dim, that their satellites, if they have any, are likely to remain undiscovered.

§ **113.** A singular proportion exists in the planetary distances. The interval between any two is twice as great as the inferior interval, and only one half the superior interval. The interval between the orbits of Venus and the Earth is twice the distance between Venus and Mercury, and one half of that between the Earth and Mars.

Let *a* represent the distance of Mercury from the Sun, *b* the interval between Mercury and Venus, then

$a =$	37,000,000=the	distance of	Mercury.
$a+2^0b=$	68,000,000=	"	Venus.
$a+2^1b=$	95,000,000=	"	Earth.
$a+2^2b=$	142,000,000=	"	Mars.
$a+2^3b=$	288,000,000=	"	Asteroids.
$a+2^4b=$	485,000,000=	"	Jupiter.
$a+2^5b=$	890,000,000=	"	Saturn.
$a+2^6b=$	1,843,000,000=	"	Uranus.

But the distance of Neptune, the outermost known planet, differs widely from this law.

§ **114.** The relative sizes and distances of the sun and planets cannot be accurately represented by orreries. We may form an idea of their proportions by drawing a circle representing the sun's disc, and placing within it all the primaries and secondaries of their true proportional size.

The following calculations give the true proportions of the size and distances of the planets. It is better, by stretching the conceptive faculty, to imagine these and other celestial facts, than by machinery to get a partial or false impression.

No machine can *truly* represent the motions of the heavenly bodies, because they are acted upon by so many influences at once, so that their motions are not what they would be in obedience to a single law. Machines are liable to another objection; they act by contact, by impulse, by visible material connection. In the heavens we see no such connection or contact between the parts which act on one another. The student is advised therefore to imagine to himself the relative sizes and distances and motions now to be described, in order to strengthen his mind to grasp the more complex phenomena to be introduced hereafter. An humble part of imagination, the conceptive faculty, from want of encouragement is apt to die out before those studies are pursued in which it is most needed. Those who can bring vividly before the mind the motion and mutual actions of the heavenly bodies, find Astronomy easy, enjoy and realize its grandeur; those whose conceptions are dull and perplexed, may hear its sublimest truths in vain. This faculty, the power of conceiving figures and motions is in the power of all.

§ **115.** Let the earth be represented by a globe 1½ inches in diameter, the proportionate diameters of the other planets would be as follows:—

Mercury, ...	½ of an inch.
Venus, ...	1 inch.
Mars, ...	¾ of an inch.
Jupiter, ...	16 inches.
Saturn, ...	15 "
Uranus, ...	6 "
Neptune, ...	5 "

While the Sun would be represented by a massive globe whose diameter would be 14 feet.

Preserving the same scale for the distances of the planets from the Sun they would be:—

Mercury, ...	190 yards.
Venus, ...	360 "
Earth, ...	500 "
Mars, ...	760 "
Asteroids, ...	1,400 "
Jupiter, ...	2,600 "
Saturn, ...	4,800 "
Uranus, ...	9,500 "
Neptune, ...	15,000 yards, or 8½ miles.

Assuming the rate of motion of a common ball to be in round numbers 1,000 miles an hour, or about 17 miles a minute, to traverse the distance between the Sun and Mercury would require its uniform flight during about 4¼ years; from the Sun to Venus, 8 years; to the Earth, nearly 11 years; to Mars, 16 years; to the Asteroids, 28 years; to Jupiter, 59 years; to Saturn, 102 years; to Uranus, 200 years; and to Neptune, 300 years.

§ **116.** Since the Sun is 330,000 times as massive as the Earth, bodies are drawn to him at equal distances with a force 330,000 times as great as terrestrial gravitation. Yet so distant from him are the planets that if let fall from their mean distances towards the Sun they would occupy in falling to it the following times:—

	Days.	Hours.
Mercury,	15	13
Venus,	39	17
The Earth,	64	13
Mars,	121	10
Vesta,	235	0
Juno,	281	5
Ceres,	297	6
Pallas,	301	4
Jupiter,	765	19
Saturn,	1,901	0
Uranus,	5,425	0
Neptune,	10,600	0
The Moon would fall to the Earth in	4	20

§ **117.** The Sun, and all the planets yet accurately examined, have spheroidal forms, differing from one another only in being more or less flattened at the poles. They have a motion of rotation round a fixed axis, from west to east, which is also common to some at least of the satellites. The planets and satellites have likewise a motion of revolution round the sun in the same direction as their rotation from west to east. The axis of revolution, like that of rotation, is always parallel to itself. In different planets it points toward different quarters of the heavens. The paths described by the planets in their revolutions are ellipses, differing only in being more or less elongated. The paths described by the satellites are ellipses in respect to their primaries, but as at the same time they move round the sun also, their real path is compounded of these two motions.

The planets nearest the sun revolve more rapidly than those more distant. They also rotate more slowly. Thus while their year is shorter, their day is longer, than that of more distant planets. Rotation has not been ascertained of the two outer planets, though it may be presumed. The following table shows the periods of rotation and of revolution which have been ascertained.

	Day and Night.		Time of revolution.		
	Hours,	Minutes.	Years.	Months.	Days.
Mercury,	24	6	0	2	28
Venus,	23	21	0	7	15
Earth,	23	56	1	0	$\frac{1}{4}$
Mars,	24	39	1	10	21
Asteroids,	unknown		4	0	21
Jupiter,	9	56	11	10	17
Saturn,	10	29	29	5	24
Urunus,	9	30	84	0	37
Neptune,	unknown		164	0	

§ **118.** Not only is the rotation of the more distant orbs more rapid, but the orbs themselves are larger. Their equatorial particles therefore describe very large circles with extreme rapidity, moving much faster than the equatorial particles of the earth. The rate at which a man at the equator of the earth rotates is $8,000\times3=$ 24,000 miles divided by $24,=1,000$ miles an hour. An inhabitant of Jupiter rotates $90,000\times3=270,000$ miles in 10 of our hours, or 27,000 miles an hour.

The day and year of the planets have here been reckoned in days and years of the earth. Since the day of the outer planets is much shorter than ours, its year contains many more of its own days than it would of ours. Jupiter's year contains 10,000 of his own days. Saturn's years 30,000 of his days.

§ **119.** The ellipses described by the planets approach very nearly to circles. The excentricities in parts of the semi-axes are :—

Mercury,	0.2055149	Ceres,	0.0784390
Venus,	0.0068607	Pallas,	0.2416480
Earth,	0.0167836	Jupiter,	0.0481621
Mars,	0.0933070	Saturn,	0.0561505
Vesta,	0.0891300	Uranus,	0.0466794
Juno,	0.2578480	Neptune,	0.0087195

§ **120.** The velocities in miles per second with which the planets move in their orbits are as follows:—

Mercury,	30 miles.	Ceres,	11 miles.
Venus,	23 "	Pallas,	11 "
Earth,	19 "	Jupiter,	8 "
Mars,	15 "	Saturn,	6 "
Vesta,	13 "	Uranus,	4 "
Juno,	12 "	Neptune,	3 "

§ **121.** We have seen that the orbits of the planets differ as to size and form, they also are performed in different planes. These planes do not vary much from the plane of the sun's equator. If we imagine the plane of the sun's equator to be extended throughout the solar system, the planets and moons will in one part of their orbits be on one side of this plane, in the other part on the other side of it.

The axis of rotation in all planets which have been closely observed, does not coincide with the axis of revolution; consequently the plane of each planet's rotation differs from that of its revolution.

The plane of the sun's rotation also is inclined to its orbit. The motion of revolution of the sun is a motion forced upon it by the planets. As they move round the common centre of gravity the sun cannot remain stationary. By virtue of his greater mass he remains near the centre of the system, and compels planets and moons to circle round him. But in return, their united imfluence forces on him a slight irregular motion round a point which always lies near his own surface.

This revolution is performed in an imaginary plane, called the fixed ecliptic, determinable from the velocities and the masses of the planets, which like the centre of inertia, never changes its position, on account of the mutual actions of the bodies of the system. This plane of inertia is called the fixed ecliptic. Its situation is nearly half way between the orbits of Jupiter and Saturn. It is inclined at a small angle only to the plane of the earth's orbit, which is called the earth's ecliptic.

§ **122.** It is more convenient to compare the planes of the planets with the plane of the earth's orbit or the ecliptic, than with the plane of the sun's equator. The following table gives the inclination of each planet's orbit to the ecliptic, and the inclination of the planet's equator to its orbit:—

	Inclination of orbit to ecliptic.	Inclination of planet's equator to its orbit.
Mercury,	7°	"
Venus,	3° 24′	75°
Earth,	0° 00′	23° 28′
Mars,	1° 51′	29° 30′
Asteroids,	7° 8′, 13° 5′, 10° 37′, 34° 37′,	"
Jupiter,	1° 19′	3° 5′
Saturn,	2° 30′	27°
Uranus,	0° 46′	"
Neptune,	1° 47′	"
Sun,		7° 10′
Moon,	5° 8′	0°

The satellites of Jupiter are inclined to Jupiter's orbit:

The First,	3° 5′ 30″
The Fourth,	2° 58′ 48″

The orbits of the seven interior satellites of Saturn are nearly circular, and very nearly in the plane of the ring. That of the eighth is considerably inclined to the rest, and approaches nearer to coincidence with the ecliptic.

The orbits of the six satellites of Uranus are inclined about 78° 58′ to the ecliptic, and their motion is retrograde. The orbits appear to be nearly circles.

§ **123.** The rapid succession of day and night in the remote planets, may, by the activity which it excites, modify the torpidity caused by the length of the year and by the great distance from the sun. At Mercury the sun shines with seven times the intensity experienced on earth, and at Uranus his radiation is at least three hundred and thirty times weaker than with us. Between Mercury and Uranus there is an actual disproportion in the quantity of solar light shed upon them of upwards of two thousand to one. Yet Uranus is not obscure; it receives as much light at noon-day as if nearly one thousand of our pale moons were shining in its sky. Neptune, in a given space, receives about $\frac{1}{1000}$ part of the light which the earth receives.

§ **124.** Some idea may be formed of the effect its greater distance has upon the climate of each planet, by considering the sun's apparent size as viewed from each planet.

The sun's	diameter	seen from	Mercury,	would be	85′
"	"	"	Venus,	"	46′
"	"	"	Earth,	"	32′
"	"	"	Mars,	"	21′
"	"	"	Vesta,	"	13′
"	"	"	Juno,	"	12′
"	"	"	Ceres,	"	11′
"	"	"	Pallas,	"	11′
"	"	"	Jupiter,	"	6′
"	"	"	Saturn,	"	3′
"	"	"	Uranus,	"	1′
"	"	"	Neptune,	"	1′

§ **125.** How far the atmospheres of the nearest and most remote planets may modify the otherwise intense heat and cold we know not, nor what affect they may have on organic life. Creations different from those we know may people these globes, and no more perceivable by our means and senses than the animal is by the vegetable world. We do not even know the effect of our atmosphere on ourselves. We know not what baneful or what blessed influences it excludes. We are born within it, we can never lay it aside or judge how far it modifies our perception of all beyond it. But we may safely infer from analogy that the sun is, in those worlds, as in our own, the source of light, heat, growth and motion. Through his influence the winds blow, the waters inundate, the earth is clothed with verdure and prepared for the habitation of man.

§ **126.** In these remote regions we find also, as some compensation, more of those satellites which so much adorn them. Mercury, Venus and Mars know only the stars; but Jupiter has four moons, each larger than ours, constantly circling around him and varying his skies. Saturn has eight and Uranus six. Neptune is also attended by one and probably by more.

The moons of Jupiter revolve in 1d. 18hs.; 3d. 13hs.; 7d. 4hs.; 16d. 17hs.

Those of Saturn in 2d. 3hs.; 1d. 9hs.; 1d. 21hs.; 2d. 18hs.; 4d. 12hs.; 15d. 23hs.; 21d.; 80d.

Those of Uranus in 5d. 21hs.; 8d. 17hs.; 10d. 23hs.; 13d. 11hs.; 38d. 2hs.; 107d. 17hs.

All but three of these periods are shorter than our lunar month, and most of the orbits are very much larger, so that the moons display immense activity, and a rapidly changing series of phases and eclipses.

Unless they had rapid motions of their own, giving them energetic tendencies to fly off, the immense attraction of the vast globes round which they revolve, would absorb them in their mass.

§ **127.** Although the satellites are usually spoken of as revolving around their primaries, this is not strictly the truth. Each planet with its satellite perpetually keeps itself balanced on each side of the common centre of gravity, and it is this centre of gravity, which, properly speaking, moves round the sun. Thus the moon forces the earth to adjust itself at such a distance from the centre of gravity as to balance itself. Thus the path of Jupiter and of each of his moons undergoes continual modifications in order to preserve the centre of gravity. If the centre of gravity is preserved, and one of the bodies, the satellite, has a revolving motion, the earth also must slightly revolve or sway nearer and then farther from the sun than the centre of gravity. Since this is the case the moon must be each half month, alternately, nearer to and then farther from the sun than the earth; the earth therefore is each half month farther from and then nearer to the sun than the centre of gravity is.

The periods of rotation of the satellites as far as ascertained are equal to the the times of their revolution. Consequently these bodies always turn the same face to their primaries.

§ **128.** The manner in which the equal times of the moon's rotations and revolutions bring the same face always present to the earth, may be seen by moving round a centre without rotating it, a ball painted half white and half black. If its white face is turned toward the sun when in one position, and it is then moved onward without rotating, when it has performed one quarter of its revolution, only a half of the white face will be toward the centre. When it has performed half the circuit none of the white face will be toward the centre. Thus without rotation the white face cannot remain visible from the centre. But if the ball roll slowly round keeping presented to the centre as large a proportion of its equator as it passes through of its orbit in a given time, it will finish its rotation and revolution in the same time and have the same face always toward the centre.

In the same way a person who begins to ascend a circular flight of stairs with his back toward a certain wall, finds himself obliged to rotate once in the course of his ascent on reaching the top his back is toward the same wall, but it has been in every other direction during the ascent. If he tries not to rotate but keeps his back obstinately in the same direction, his face cannot always be toward the centre. The rotations and revolutions are always in the same direction from west to east with the exception of the satellites of Uranus.

§ **129.** As we know but little of the more distant satellites, and as the phenomena appear the same in all, with the exception above mentioned, we will now confine ourselves to the earth's moon.

The plane of the moon's revolution is inclined to the ecliptic 5° 8′. It moves eastward at the rate of two thousand miles an hour and completes its revolution in twenty nine days.

The axis of the moon's rotation is inclined to the pole of its orbit, and always preserves its parallelism with itself.

Beside the moon's rotation, and revolution round the

earth, it revolves round the sun, as the earth does, in a large ellipse. If it felt only the earth's attraction, it would describe an ellipse with the earth in one of the foci, returning to the same place at the end of each month. But as it performs at the same time a small ellipse round the earth and a large ellipse round the sun, the path really described in space is a compound of these two motions, it is a succesion of curves, varying in concavity, but always concave toward the sun.

§ **130.** Let us imagine two persons fastened together by a rod of a certain length which compels them to keep always at the same distance, one from another. Let them both describe a large circle round a tree, and let the smaller one at the same time go round his companion. And let the companion walk so rapidly, that the other can never actually return to his former place but is forced to move on.

We have here a rude image of the moon's motion round the sun. The earth, rushing a million and a half miles daily in her orbit, bears on the moon and straightens out the curve she would otherwise describe. In some parts of the moon's monthly orbit her course is accelerated, in others it is delayed, and the curve consequently varies in concavity.

Let us begin with the moon between the sun and earth, the new moon. During the first quarter of her orbit, the moon's revolution round the earth will tend to carry her backward in space. Her revolution round the sun tends to carry her onward. The former motion will tend to delay the latter. In the next two quarters of her monthly revolution, the two motions will coincide in direction, and the result will be greater rapidity of motion. Fig. 2d, Plate I., gives the motion of the moon for one month.

Jupiter is so large and so near to his satellites, in comparison with the sun, that the curves which they describe are different from the path described by our moon, although they go round Jupiter as the moon goes round the earth.

§ **131.** Let A, B, C, D, E, Fig. 3, Plate I, be as much of Jupiter's orbit as he describes in eighteen days; and the curves a, b, c, d, will be the paths of his four moons going round him in his progressive motion. The first satellite intersects its own path once in 42½ hours, making such loops as those in the diagram. The second satellite, moving more slowly, crosses its own path once in three days, thirteen hours, making out five loops in the time in which the first makes ten. The third satellite, moving still more slowly, comes to an angle at the end of seven days, four hours, and then describes another such curve. The fourth satellite is always progressive, making neither loops nor angles in the heavens.

Those satellites whose velocities round their primaries are greater than the velocities of their primaries in open space, make loops when nearest to the sun. This is the case with Jupiter's first and second satellites, and with Saturn's first.

But those satellites whose velocities are less than the velocities of their primary planets, move direct in their whole circumvolutions. This is the case with the third and fourth satellites of Jupiter, and with the second, third, fourth and fifth, satellites of Saturn, as well as with our moon.

As the moon turns upon her axis in precisely the same time which she takes to revolve about the earth, she has but one day or night in one of our lunar months; and as she encompasses the earth thirteen times during the earth's progress round the sun, it is manifest that a lunar year contains about thirteen lunar days.

§ **132.** There are striking differences in the relative sizes and weights of the planets. And the different proportions of their size and weight cause essential differences in their material composition.

The following table shows the known sizes and densities of the members of the solar system.

	Mercury.	Venus.	Earth.	Mars.	Jupiter.	Saturn.	Uranus.	Sun.	Moon.
Diameter in miles,	3140	7700	7916	4100	90000	76000	35000	883000	2160
Volume, that of the earth being one,	0.06	0.93	1.00	0.14	1470	887	77	1328460	0.02
Mass, that of the earth being one,	0.16	0.94	1	0.13	338	120	17	354936	0.013
Density, that of the earth being one,	2.95	.99	1	0.79	0.25	0.11	0.26	0.26	.75

In comparing the size of bodies, we must observe whether their diameters, their discs, or their solid contents are compared. The discs are to each other as the squares, and the volumes as the cubes of the diameters.

The mass of the sun bears to the mass of the earth but a small ratio compared with that which its volume bears to the volume of the earth. In judging the volume of the sun, we take the extent of the bright surface, which probably is an atmosphere. This atmosphere may be many thousands of miles deep, and of course has a low specific gravity. The density ascribed to the sun is however composed of the density of this atmosphere and of that of the sun's body, which may be very great. The sun proper, without its atmosphere, as we calculate the planets, would have a smaller size and a greater density.

§ **133.** The densities of two bodies are directly as their masses or weights, and inversely as their volumes. More frequently the density is compared with that of a globe of water of the same size, and the specific gravity is thus obtained. Thus the Sun weighs $1\frac{1}{4}$ compared with a globe of water of the same size; Mercury $17\frac{7}{10}$; Venus $5\frac{1}{5}$; Earth $4\frac{9}{10}$; Mars $3\frac{3}{10}$; Jupiter $1\frac{1}{10}$; Saturn $\frac{1}{2}$; Uranus 1.

The matter of which Mercury is made is nearly four times as heavy as that which composes our earth, while Saturn is as light as cork or deal. Mercury has the density of quick-silver; Uranus and the Earth, that of steel; Mars and the Moon, that of diamond; the Sun and Jupiter, nearly that of resin. Upon the weight or mass of each planet depends, chiefly, the weight of bodies near its surface.

§ **134.** If we suppose two planets, of equal masses, but one of one hundred times the density of the other, a man at the surface of the smaller one would weigh most, because he would be nearest the centre of gravity. If these two planets were rotating in the same time, and the man stood near the equator, the difference of weight would still be increased, because the surface which performed the largest circle would generate the most centrifugal force, and thus diminish his weight most. Weight on the surface of a planet at rest is in direct proportion to the mass of the planet, and in inverse proportion to the square of the distance from the centre. On two bodies of unequal density, but equal size, weight is greatest at the densest; on two of equal density, but unequal size, greatest at the largest. Weight on a large, not dense body, may be just equal to weight on a small dense one. If both bodies rotate, the weight at the surface of each decreases rapidly, in proportion to the rapidity of rotation. Thus the larger planets exercise less attraction on bodies at their equators, on account of great rapidity of rotation.

§ **135.** The number of particles in the attracting mass being changed, the conditions not only of inorganic but of organic creations are altered. There is a limit to the size of animals, trees, buildings, in each planet. No house can stand when made so large or so loosely that the cohesive force of its parts does not overcome the attractive force of the earth. No animals now exist on earth so large as those which reposed in the marshes of the primitive world. The sea yet has its whales, because it bears up their bodies, and thus as it were diminishes gravitation. If man, with his present organization, were transported to the surface of a body as large as the Sun, he would probably fall to pieces like a figure of smoke. If the cohesion of his body were increased, he would yet be unable to move, unless greater muscular power were granted. For the attractive force of the Sun would cause bodies to fall through 334 feet in a second, and would consequently attract man thirty times as strongly, and give him thirty times as much weight as he has here. A man of moderate size would weigh about two tons at the surface of the sun. Whereas at one of the asteroids, he would weigh but a few pounds, and would find it difficult to remain attached to the planet.

§ **136.** We have considered the Sun as the controller of the system and its motions, and as the dispenser of light and heat; we will now consider him as he clothes other bodies with light. Not only all the day-light, but all the planet and moon-light of our system comes from him. As the train revolve, both of primaries and secondaries, one half of each orb is lighted up by him. One half of each is always in light, one half in shade. When two planets, or a moon and a planet are in such a position that the light of the Sun is reflected from one to the other, the former becomes visible by reflected light. Whenever a planet or a moon passes between another planet and the Sun, the Sun's light will be cut off, at least partially, from the second planet. We will consider what interferences can arise in the solar system, and as we view them from the earth we will mention those visible from the earth, and give them the names usually applied.

A body may disappear on account of another body's coming between it and the source of its light, or on account of another body's coming between it and the spectator. In the former case, the body is really eclipsed or darkened; in the latter case, it remains illumined, but is no longer seen by us. We apply the term eclipse to both of these occurrences, though they differ widely in their nature. A real eclipse is the same viewed from all parts of the system; an apparent eclipse is only an eclipse to one particular place, sometimes only to one portion of the earth at a time.

§ **137.** Eclipses of the first kind take place when the moon passes into the earth's shadow; or when any of the satellites enter the shadow of their primaries.

If the planets were so large, or so near to one another, that the shadow cast away from the sun by an inner one could reach the surface of an outer one, the outer one would be eclipsed whenever they passed in their orbits; but their small size and mutual distance forbids this.

Partial eclipses of this kind take place whenever the satellites pass between their primaries and the sun, and cast their shadows on the discs of their primaries. On such occasions, to those portions of the planet on which the shadow falls, the sun appears eclipsed.

Eclipses of the second kind occur:—

When the moon passes between the sun and the earth, and the sun is eclipsed to the earth. When the moon passes between a fixed star or a planet and the earth, cuts off its light and occults it. Or what is less observable, when the sun occults a planet or a star. Planets sometimes, but rarely, eclipse one another.

When Mercury or Venus passes between the earth and the sun, and intercepts from our view a small portion of his disc, it then appears on the sun's disc as a little black ball, and its passage is called a transit.

In like manner the satellites of Jupiter may conceal from us a portion of Jupiter's illuminated surface, or Jupiter may conceal from us its satellite's illuminated surface. A satellite of Saturn may pass between its ring and the earth, or Saturn or his ring may occult a satellite.

Eclipses of the first kind, take place when the sun, the interposing, and the eclipsed body are in one straight line. Eclipses of the second kind take place when the observer, the interposing, and the eclipsed body are in one straight line.

§ **138.** If the various members of the solar system moved in the same plane eclipses would take place much more frequently than they now do. Mars and Venus would pass between the earth and the sun almost once in each one of their revolutions. The moon would eclipse the sun every month, and would itself pass into the earth's shadow every month. As all the planes are inclined to one another it is but rarely that the centres of any three of the bodies are in one straight line.

Bodies in the heavens are often invisible to an observer on earth, from two other causes, because they are lost in the superior light of a neighboring body, and because they are so situated that their bright surface can not send any rays to the earth. When Mercury and Venus are nearly in a straight line between the sun and earth, the greater part of their surface which is toward the centre, is dark and therefore invisible ; the only part visible is a portion of the illuminated surface. When the planets are nearly in a line with the earth and sun, but beyond the sun, their bright surface is toward us but is lost in his rays.

When the moon is between the earth and the sun, her bright side is toward the sun and she is invisible for some hours. When the moon is beyond the earth her whole illuminated hemisphere is visible from the earth. In all situations between these two, a portion of her illuminated hemisphere is visible from the earth.

Of course only opaque bodies, such as shine with reflected light can become invisible from this cause. The moon's dark side might perhaps be visible here as it is said to be in Syria, and as we see the old moon in the new moon's arms, if it were not when new so near the sun.

§ **139.** We have now brought before us one by one all the circumstances of the solar system. We have imagined the sun balancing by his mass the revolving planets. These with their satellites, each in its particular plane, and with its own velocity, never resting, moving from the beginning of its creation, and turning on its axis, all obediently circle round the sun. We have seen each with its pole of rotation invariably pointed toward the same star, wherever the motion of revolution bears it on. One half of each orb is bathed in light, the other plunged in darkness. All are always in the starlight, but the stars are seen by none till it has turned into its own shadow.

For three thousand millions of miles on every side the obedient orbs recognize the central power. Beyond this may lie other subjects, but their reflected light is too dim to attract our eyes. We may hereafter learn their existance from the perturbations of Uranus and Neptune.

We have found that the members of the solar system differ in many important particulars; in distance from the sun; in times of rotation, and of revolution; in mass; in density; in degree of compression at the poles; in the planes and ellipticities of their orbits; and in the inclination of the plane of the rotation to that of the revolution of each planet. Most of these circumstances must affect greatly the physical condition of the surface of each planet. Many minor causes also, as the nature of the atmosphere or its absence, the presence of water and many unknown causes doubtless introduce still greater variety.

§ **140.** Some of these differences observed among the planets appear to follow a law, and doubtless a law always exists though undiscernible by us. The planets within the Asteroids are of more moderate size, are more dense, rotate round their axes more slowly, and in nearly equal periods, and are less compressed at the pole than the planets beyond the Asteroids, and with one exception, are without satellites. The outer planets are of much greater magnitude, are less dense, more than twice as rapid in their rotation, more compressed at their poles, and possess all but one of the satellites of the system. These remarks cannot however be applied strictly to each planet. Nor are there any constant relations between the distances of the planets from the sun, and their absolute magnitudes, densities, times of rotation, eccentricities, and inclinations of orbit and of axis. Neither in size nor density is there any regular succession as we go from the sun. The time of rotation decreases on the whole with the increasing solar distance, yet it is greater in Mars than in the earth, and in Saturn than in Jupiter. Juno, Pallas and Mercury have the greatest eccentricity; and Venus and the earth which come between them have the least. Nor is there more regularity in the inclination of the orbits, or the position of their axes of rotation relatively to their orbits; though on the whole those planets which have the most elongated orbits, have their orbits most inclined to the ecliptic. The orbits of the different planets are elongated in different directions; the position of the major axis of each orbit is not however invariable.

§ **141**. In the relative size of the moons and primaries no law is discoverable. The earth's moon is of great relative magnitude, its diameter being to that of the earth, as one to four, whereas the diameter of the largest of all known satellites, the sixth of Saturn, is but one seventeenth, and and that of Jupiter's largest satellite is but one twenty-sixth part of the respective diameters of the planets round which they revolve.

The density of the moon is three fourths that of the earth, while the second satellite of Jupiter appears to be actually more dense than the great planet round which it revolves.

The satellites of Saturn offer the greatest contrasts both of absolute magnitude, and of distance from the central planet. The sixth satellite is but little smaller than Mars, (whose diameter is twice that of our moon) while the recently discovered satellite is one of the smallest bodies in the solar system.

The absolute distance of a satellite from its primary is

greatest in case of the outermost satellite of Saturn. It is above two millions of geographical miles, or ten times the distance of our moon from the earth. The satellite which is nearest to its planet is undoubtedly the innermost of Saturn, and it offers the only example of a period of revolution of less than twenty four hours. Its distance from the centre of the planet is eighty thousand and eighty eight miles; from the surface of the planet it is forty seven thousand four hundred and eighty miles.

§ **142.** If we estimate distances not in absolute measure but in radii of the primary planets, we find that the nearest of Jupiter's satellites, which in absolute distance is twenty six thousand miles farther from the centre of that planet than our moon is from the earth, is only six radii of Jupiter from its centre, while our moon is distant from us fully sixty and a half radii of the earth.

Even the law, mentioned above, of the distances of the planets from the sun is not numerically exact for the distances between Mercury, Venus and the earth, and is violated in the case of Neptune. Even allowing this law to have no exceptions it is one found only by observation; we have no idea on what principle it is founded, nor how it acts.

But there are circumstances in the form and motions of the planets whose principle and immediate cause are known to us. These are of the deepest interest because they throw light on the past condition of the planets. The spheroidal form of the sun and planets, their two motions in the same direction, the near coincidence of the planes of their revolution with the plane of the sun's equator, give us a hint as to what forces presided at their birth.

CHAPTER VII.

METEORS AND THE ZODIACAL LIGHT.

Appearance and number of Meteors; their composition and size. Meteoric showers; their supposed origin. The Zodiacal Light; its appearance; different theories of its nature; its possible connection with Meteoric showers.

§ **143.** Thus far we have dealt with facts. All we have learned has been from observation and reasoning. If we would go farther and inquire what circumstances exerted the forces we have been tracing, we enter on the domain of theory. A theory which would explain the formation of the system must however include all its members, and there are some members of our solar system which we have not yet described. Let us make ourselves acquainted with these, and then we shall be prepared to include in one view, all the forms which matter, to our eyes, ever assumes.

There are two more bodies or classes of bodies which decidedly belong to our system; the Zodiacal Light and Shooting Stars or Meteors; the latter have even been claimed as belonging to the earth's atmosphere. Beside these are ether, which perhaps is common to our system and to the rest of space, of which we know nothing except that it is unlike every other form of matter, and comets, which we can scarcely claim as belonging exclusively to our system, but which exhibit matter under very peculiar and interesting conditions.

Having studied these objects and also the physical state of the nearer planets, we shall have the slight data on which all theories of the formation and former state of the universe are founded.

§ **144.** Shooting Stars have a particular interest: they are the only visitants from other worlds which ever reach the surface of our earth; the only foreign matter which enters the earth's atmosphere and may be touched and examined. They give us therefore our only intelligence of the physical composition of the rest of the solar system.

We become conscious of the presence of most members of the solar system either by the light they send us, or by the attraction they exercise on us as we pass them in space. We know the presence of comets because they send us light, but we do not know it from their attraction, because if they exercise any on us it is so slight as to be imperceptible.

Now it is possible that, in the immense fields of space through which we journey, there may be other travellers, whose minute size prevents our detecting them either by sight or through gravity. When we pass near these small bodies we may have no intimation of their presence. But if they enter our atmosphere, drawn by the earth's attraction, they may become luminous, and either be dissipated in the upper regions of the air, or fall to the ground. They would thus present to us the phenomena of meteors.

§ **145.** The large meteors are called globes or fireballs. The small meteors exhibit only a bright path or line, and are called Shooting Stars. These balls and stars often appear together. In general they have the same hue as the fixed stars. Their color sometimes becomes yellow, and sometimes blue or green. The trains which they leave behind them are not smoky, but rather like a shower of sparks. Sometimes the star breaks into fragments which form a continuation of the train, and which vanish almost as soon as that. When they break they sometimes let fall stony fragments, covered with a distinct shining black crust, such as our ovens could not produce. Though considerably heated they are not incandescent. They sometimes appear to have been softened, but never to have been melted during their passage through the air. Wherever they have been collected, in all periods of time, and in all parts of the earth, they resemble one another in their form, in the nature of their crust and in their chemical composition. About one third of the elementary substances which compose our globe have been found in them.

They are generally composed of metals, among which nickel, cobalt and virgin iron are the most common, or they are clayey and contain crystals. They fall with a force which causes them to sink from ten to fifteen feet into the earth. Their form proves that they are fragments. None have been known to fall more than seven or seven and a half feet in diameter. They are usually of much smaller size, and many seem never to reach the earth, and either to have no perceptible mass, or to be entirely dissipated in the atmosphere. These may perhaps have small nuclei surrounded by inflammable vapors or gases. Some of the fire balls which appear the largest may be of this kind. The apparent size and brightness seem to have no connexion with the size of the fragments let fall.

§ **146.** Meteors are visible in great numbers and in all parts of the heavens. A register kept from July 1841 to February 1845 gives five thousand three hundred and two Shooting Stars, observed in one thousand and fifty four hours, from one observatory. Among these were eight globes, and eighty Shooting Stars of the first magnitude. Whence it follows that an observer would see one globe a week, and one falling star of the first magnitude, each night of eleven hours.

Single meteors such as we have described appear in all quarters of the heavens. They vary as to swiftness and as to height; some being not more than sixteen and others one hundred and forty miles high. The largest appear to have the greatest altitude, and only the smaller ones appear to come within twenty or even forty miles of the earth. The motion in all cases is not in the same direction; the prevailing direction is from north east to south west, contrary to the motion of the earth in its orbit. This direction of the motion is particularly observed in those meteors which fall in showers. They come from the same point during the whole continuance of a shower, which proves their independence of the earth's rotation, and consequently that they come from without our atmosphere. This fact and the periodical recurrence of the showers has given to meteors a new importance. It has made improbable the before received theories concerning their origin; unless indeed we suppose that there may be several kinds produced by different causes. It is not well ascertained of what importance these periodical showers are, but they are too striking and peculiar to be overlooked, and they may throw light on other unexplained phenomena.

§ **147.** Single meteors or even showers of stars appearing irregularly have been explained by supposing them to be gaseous substances condensed in the upper regions of the atmosphere, perhaps by the same agency which condensed the earth. Their composition, of metals found also in the earth, favored this hypothesis.

The hypothesis of their lunar origin has also found some believers. The great size and height and the crater-like form of the lunar mountains led to the supposition that they were extremely active. A body projected from a volcano in the moon, with a velocity of about eight thousand five hundred feet a second, would not fall back on the lunar surface, but would recede from it indefinitely. In order to reach the earth it would require a velocity of only eight thousand three hundred feet. Such a velocity, which is only about four or five times that of a cannon ball, is quite conceivable. But the extraordinary exhibitions of 1799 and 1833 are quite irreconcilable with a lunar origin.

To be satisfactory, a theory must explain not only their coming in showers, but the periodical recurrence of the showers in the months of August and November.

It is only within seventy years that attention has been directed to this subject and though on looking back some traces of periodicity have been found, they are scarcely sufficient to establish their periodicity as a law. The principal displays have been in 1799, 1832, 1833 and 1844.

On the 11th of November, 1799, thousands of shooting stars were observed by Humboldt, at Cumana; and on the same night by different persons, over the whole continent of America, from Brazil to Labrador, and also in Germany. In 1832 they were seen over the whole of the north of Europe, and in 1833 the wonderful display took place in

North America, which has been so well described by Professor Olmsted, and which first established the importance of the subject. He thus describes the great meteoric shower of the 13th of November, of that year.

§ **148.** "On that morning, from 2 o'clock until broad daylight, the sky being perfectly serene and cloudless, the whole heavens were lighted with a magnificent display of celestial fire works. At times, the air was filled with streaks of light occasioned by fiery particles darting down so swiftly as to leave the impression of their light on the eye, (like a match ignited and whirled before the face,) and drifting to the north west like flakes of snow driven by the wind; while, at short intervals, balls of fire, varying in size from minute points to bodies larger than Jupiter and Venus, and in a few instances, as large as the full moon, descended more slowly along the arch of the sky, often leaving after them long trains of light, which were, in some instances, variegated with different prismatic colors.

On tracing back the lines of direction in which the meteors moved, it was found that they all appeared to radiate from the same point, which was situated near one of the stars, (Gamma Leonis) of the Sickle, in the constellation Leo; and in every repetition of the meteoric shower of November, the radiant point has occupied nearly the same situation.

This shower pervaded nearly the whole of North America, having appeared in almost equal splendor, from the British Possessions on the north, to the West India Islands and Mexico on the south, and from sixty one degrees of longitude east of the American coast, quite to the Pacific Ocean on the west. Throughout this immense region, the direction was nearly the same. The meteors began to attract attention by their unusual frequency and brilliancy, from *nine to twelve*, in the evening; were most striking in their appearance from *two to four;* arrived at their maximum, in many places, about *four o'clock;* and continued until rendered invisible by the light of day. The meteors moved in right lines, or in such apparent curves, as, upon optical principles, can be resolved into right lines. Their general tendency was toward the north-west, although, by the effect of perspective, they appeared to move in all directions.

§ **149.** It is considered as established that the meteors had their *origin* beyond the limits of the atmosphere, having descended to us from some body existing in space independent of the earth; that they consisted of exceedingly light combustible matter; that they moved with very great velocities, amounting in some instances to not less than fourteen miles per second; that some of them were bodies of large size, probably several hundred feet in diameter; that when they entered the atmosphere, they rapidly and powerfully condensed the air before them, and thus elicited the heat which set them on fire, as a spark is sometimes evolved by condensing air suddenly by a piston and cylinder; and that they were consumed and dissolved into small clouds at the height of about thirty miles above the earth."

Professor Olmsted referred this periodical return to astronomical causes and predicted its return at the same season, in future years. It was visible in different parts of the earth every year until 1839, and since then it has ceased altogether.

§ **150.** The following is Professor Olmsted's reasoning, and his theory.

Since the earth fell in with the meteoric body in the same part of its orbit, several years in succession, the body must either have remained there during a year, or it must itself have had a revolution round the sun. No body can remain stationary in the planetary spaces, or it would be drawn either into some nearer body or into the sun. The body whence meteors fall must therefore have revolved either in a year or some aliquot part of a year, or it could not have come in contact with the earth so many successive years. If it revolves in an elliptic orbit it will some years encounter the earth and other years pass at a distance from it. This may explain the absence of the showers for several years. The meteoric body is too small to be seen. It probably consists of myriads of planetoids, which, for all we know, may fill the planetary space. They may circulate about the sun, generally in groups or zones, and two of the zones may intersect that part of the earth's orbit through which it passes in August and November. When the earth encounters a thin portion of the zone the showers are scanty and if the intervals in the zone are wide, only scattered meteors will be visible, as on ordinary nights.

This zone of planetoids may be as old as the larger planets, and may rank as an important portion of the system. It may consist of those portions of matter which were not sufficiently near one another to be attracted into one mass.

It is possible that this revolving zone may be composed not of solid bodies, but of nebulous matter like the tails of comets. We can more easily understand the disappearance of nebulous than of solid matter in our atmosphere, and a very large proportion of meteors never touch the earth.

It has also been suggested that meteors may have their origin in the zodiacal light, a phenomenon hereafter to be described. Since the plane of this nebulous substance is not parallel to the ecliptic, the earth might pass through it at one season, and be remote from it another. But this does not account for the appearance of shooting stars at all seasons of the year.

The interruption of these phenomena may be caused by a motion of the nodes of the stream of aerolites, so that what has at former periods been so striking, and what has been repeated in our own times, will again recur after an interval.

§ **151.** The zodiacal light and meteors, although very unlike one another in appearance, may perhaps arise from similar causes. The zodiacal light is a pale cone of light projected from the sun after the evening and before the morning twilight. It is almost constantly visible in the torrid zone, but in northern temperate regions, is only distinctly visible in the beginning of spring, after the evening twilight, and at the end of autumn before the commencement of the morning twilight. Its light resembles that of a comet. The faintest stars may be seen through it. It is less bright than the milky way, with ill defined edges, scarcely to be distinguished from twilight. Humboldt describes it in 10° latitude as appearing very regularly about an hour after sunset. Before this, even if the night was perfectly dark, no trace of it could be seen. Then it suddenly became visible, extending in great brightness and beauty between Aldebaran and the Pleiades and attaining an altitude of 39°. Long narrow clouds appeared low down on the horizon, as if in front of a golden curtain, while bright tints played on the upper clouds. The light diffused in that part of the heavens appeared almost to equal that of the moon in her first quarter. When its brightness was greatest a mild reflected glow was visible in the east. Towards ten o'clock it became very faint, and at midnight only a trace of it remained.

§ **152.** Its figure agrees with that of a spheroid seen in profile. It has the sun for its base, and its axis lies nearly in the direction of the zodiac whence it takes its name. It also lies very nearly in the plane of the sun's equator.

As the sun's equator is differently inclined to the horizon, on account of the different positions of the sun in the ecliptic, the zodiacal light inclines with it, and is in a great measure concealed beneath the horizon; or at least its lustre is diminished by vapors. In the vernal equinox the arc of the ecliptic which the sun is about to enter is more elevated above the horizon of a place in north latitude than the equator is. The zodiacal light is then elevated above the equator by all the obliquity of the ecliptic; no other position is so favorable for our climate. In the summer solstice the arc of the ecliptic, and consequently the luminous cone, is parallel to the equator, and therefore much more inclined to the horizon of places in north latitude than in the spring.

The apparent angular distance of its vertex from the sun varies according to circumstances, from 40° to 90°; and the breadth of its base perpendicular to its axis varies from 8° to 30°. It must involve Mercury and sometimes Venus and the Earth, and if it were not extremely rare, would produce some disturbance in their motion, but in fact it does not appear to impede the progress even of the tails of comets.

As to its probable composition we must choose between the supposition of its being purely nebulous, or loaded with the tails of millions of comets; or of its consisting of a stream of countless planetoids or meteors, too small to be seen separately, but able from their numbers to give out a faint light. This latter hypothesis has this advantage, that it resembles the cause assigned for periodic shooting stars.

CHAPTER VIII.

COMETS.

The number of recorded Comets. Variety in their motions and appearance. Their immense size. Description of a Comet. The tails of Comets. Bessel's Theory of their formation. Halley's Comet. Biela's and Encke's. Their resistance by the ether. The mutual influence of Comets and Planets. Mass of the Comet of 1770. The probable effect of a collision with a Comet.

§ **153.** Comets form a class of bodies entirely distinct from the fixed stars and from the planets, whether we regard the character of their movements or their physical constitution. They receive their name from their hairy appearance, caused by the coma or atmosphere which surrounds them. Of the number of comets it is impossible to speak with certainty. Many comets on their nearest approach to the sun are too distant to be seen from the earth; many may not have reached their perihelion within the recorded experience of man; many may be invisible from their diminutive size; many can be seen only from the south side of the equator, where there are but few means of observation; many, though on the north side of the equator, rise above the horizon only during the day; many pass unnoticed, owing to cloudy weather. Several have however been seen so bright as to be visible in the day time, even at noon and in bright sunshine; and there is one instance on record of a very large one observed near the sun, when eclipsed, in the year 60 before Christ. The number of comets which enter our system must amount to many thousands; more than six hundred have been actually observed, and the orbits of between one and two hundred have been calculated.

They come from every region of the heavens, and move in every variety of plane. Some move in the same direction with the planets; others in the opposite direction. Some of them remain in sight for a few days only, others for many months; some move with extreme slowness, others with extraordinary velocity. Not unfrequently the two extremes of apparent speed are exhibited by the same body, in different parts of its course.

§ **154.** Not only does a comet vary in its physical appearance and its speed in different parts of one course, but it sometimes presents on its return, an appearance so different as to be scarcely recognizable. Its size and splendor are sometimes so much diminished that it is difficult to identify it.

We must not however suppose that all the apparent changes in the tails of comets are real. Many of them are owing to the state of our atmosphere, as is proved by the same comet's appearing of different brilliancy and extent, in different parts of the globe. Our atmosphere is a coarse medium, through which to view objects so delicate.

Another circumstance which makes it difficult to identify a comet is, that their orbits are liable to change after they enter the solar system, owing to the attraction of the planets. The orbit of a comet is, however, more to be relied on, as a test of its identity, than its physical appearance, if the changes in its orbit can be accounted for by the influence of any member of the planetary system.

§ **155.** We will now consider, in detail, the physical constitution of comets, their motions and the influence they may impart to, or receive from, the members of the solar system.

Comets are the most voluminous and at the same time the lightest bodies of our system.

The tails of some of the largest have extended over a

distance of from thirty to forty million leagues, a length much exceeding the interval between the sun and the earth. At the same time their weight is so slight that not one has disturbed a planet or a satellite, in the slightest perceptible degree. It follows that matter so exceedingly diffused must be transparent. Stars of the sixteenth and seventeenth magnitude, which the slightest fog would conceal, may be seen through their substance; and yet their light passes through thousands of miles of the body of the comet.

It has been doubted whether comets shine with a light of their own or by reflected light like planets. As they present no phases like the moon, it has been supposed that they originate light. They may however reflect it throughout their whole substance like the light clouds of our atmosphere, which often appear soaked in light. The fact that part of their light is polarized makes it certain that at least part of their light is reflected.

§ **156.** Comets consist of a large ill defined mass of cloudy luminous matter, usually increasing in brilliancy toward the centre. This central portion is called the nucleus of the body. The nucleus and coma belong to comets in all parts of their orbit. The tail and head are developed as they approach the sun.

The *Nucleus* of the comet is generally to be distinguished by its forming a comparatively bright point in the centre of the head. In most instances it has the appearance of a solid body, and frequently subtends an angle capable of telescopic measurement. It is usually enveloped in a dense nebulous stratum, called the *coma*. This stratum so frequently renders the edge of the nucleus indistinct, that it is extremely difficult to ascertain its diameter with any precision.

But though comets, in general, possess this nucleus or body, there are many of them in which it seems to be entirely wanting, and which present only a nebulous mass, having a gradual condensation towards the centre. Apparently, there is a regular gradation of Comets, from such as are composed merely of a gaseous or vapoury medium, to those which, by the mutual attraction and consolidation of their nebulous particles, have at length acquired a consistent nucleus. In the small comet of 1804, for example, no solid body could be discovered; it seemed to consist entirely of vapours. A star of the sixth magnitude could be distinguished through the very centre of the comet of 1796; and Herschel asserts a similar fact with regard to that of 1795. Through the comet of 1802 a star of the tenth magnitude, could be observed, with hardly any diminution of its light. The second comet which appeared in 1798, was estimated to have a nucleus of twenty-seven miles in diameter. The nucleus of the comet seen in December, 1805, was computed to be thirty miles in diameter. The comet of 1799 had a nucleus three hundred and seventy three miles in diameter. The first comet of 1811 had one four hundred and twenty-eight miles in diameter; and the second comet of that year was observed to possess a nucleus of prodigious size; from Herschel's observations it was no less than two thousand six hundred and thirty-seven miles in diameter, or one third of the earth's diameter.

§ **157.** In all comets there is an envelope of light, which in some cases seems to be united with the nucleus. This envelope almost never surrounds the nucleus, but forms a sort of hemispherical cap on the side next the sun, and then diverges into two brilliant streams on the opposite side, giving rise to the singular and well known phenomenon of the comet's tail. Some comets are furnished with several of these singular appendages. That of 1744 had six, which spread out like an enormous fan, extending to a distance of nearly 30°.

Many of the brightest comets, however, have been observed with short and faint tails, and not a few have been entirely without them, and in these cases the whole nucleus presents only a globular mass of nebulosity. This shining envelope is supposed to be of the same nature as the stratum immediately contiguous to the nucleus, viz: matter raised from the surface by the action of the sun's heat, and converted into a state of high attenuation. It is found to vary considerably, as well in its own thickness as in its distance from the nucleus. In the comet of 1811, for example, the depth of the envelope at one time amounted to twenty-five thousand miles, and its distance from the centre of the nucleus was found to be thirty thousand miles. In the comet of 1807, the depth was thirty thousand miles. The small comet of 1804, which seemed to have no solid part at all, presented a mass of nebulosity of about five thousand miles in diameter.

Herschel supposes that a very elastic and transparent medium surrounds comets like an atmosphere, in which, when the cometic matter has become sufficiently rarefied by the solar heat, it rises to a certain elevation, and remains there suspended. The transparency of this atmosphere is proved by the appearance of small stars through it. Its elasticity may be inferred from the circular form it always assumes.

§ **158.** That the atmosphere of comets must be of very considerable extent, is evident both from the great depth of the envelope, and from the not unfrequent occurrence of several of these nebulous envelopes one above another, all of which must necessarily be suspended in the same buoyant medium. As any matter suspended in such a medium must have a density inversely proportional to its height, it would follow that the outermost of these envelopes should not be so bright as those nearer to the nucleus; and this inference is fully verified by observation.

The extraordinary size of the atmospheres is owing to the slightness of the attraction of the exceedingly small central mass. If the earth, retaining its present size, were reduced, by any internal change, to one thousandth part its present mass, the atmosphere would expand to more than a thousand times its present bulk.

These nebulous envelopes are thus formed. When the comet is approaching the sun the nebulous matter suspended in its atmosphere is made to rise higher by the increasing energy of the sun's heat. For the same reason, after one envelope has risen to a considerable height, so much matter may subsequently be detached from the nucleus as to constitute a second, which, from being more dense than the first, will occupy a lower situation in the atmosphere.

In like manner a third and fourth envelope may be successively formed.

Thus the comet of 1744, which at its perihelion approached the sun to within one fifth the distance of the earth, had, about three weeks previous to its perihelion passage, a double envelope, and on the seventh or eighth day after the passage had acquired a third. As the great comet of 1811 receded from the sun, its envelope, losing its high degree of attenuation, at length subsided altogether upon the nucleus.

As the tails are much more striking after the comet has passed the sun, it has been suggested that the nebulous matter when exceedingly excited, may be invisible as steam is when it first issues, and afterward being precipitated, it becomes visible.

§ **159.** The tail is only a continuation of the nebulous envelope, which, after nearly encompassing the hemisphere of the nucleus next the sun, diverges to a greater or less extent in an opposite direction. The tail is sometimes wanting, and sometimes is forty or even one hundred millions of miles in length. The tail is always of a conical shape, the apex being the hemispherical envelope, and the base generally ten or twelve times as broad as the diameter of the nucleus.

As the whole envelope is equally exposed to the action of the sun, (which in some way or other produces the tail,) all the parts are caused by impulsion to assume the shape of a conoid, and thus the tail is hollow. Therefore the sides or edges of the tail have usually the appearance of two brilliant streams, the space between them appearing to contain a much less quantity of nebulous matter. For as the line of vision traverses a greater number of luminous particles at the sides, where that line is a tangent to the cone, than toward the middle, where the line of vision is more perpendicular to the envelope, there is a much greater quantity of light at the sides than at any other point. For the same reason, the top of the hemispherical cap, or that part nearest the sun, is generally more brilliant than any other point.

Another fact proves at once the hollowness and the conical form of the tail. In whatever position comets are placed, and they are frequently observed during a course of 180° round the sun, they constantly present the same appearance, as to the shape of the tail, and the superior brilliancy of its edges.

§ **160.** The length of the tails of comets has sometimes been enormous. That of 1618 is said to have been 104° in length; that of 1680, immediately after its perihelion passage, was twenty million leagues, and occupied only two days in its emission from the comet's body; a decisive proof of its being driven forward by some force, the origin of which, to judge from the direction of the tail, must be sought in the sun itself. The diameter of the head of the comet of 1843 exceeded one hundred thousand miles. The breadth of the tail in some places was eight hundred thousand, while the extent could not be less than one hundred and seventy million miles, or nearly equal to the diameter of the earth's orbit.

It is hardly possible that matter once projected to distances so enormous should ever be collected again by the feeble attraction of such a body as a comet.

Biela's comet, on its return in 1846, exhibited the astonishing phenomenon of a double comet. It may have been originally double, or may have become so since its last appearance, when it was seen undivided, the two portions subsequently journeying along side by side, in orbits slightly differing from one another, and apparently quite undisturbed by any mutual attraction.

§ **161.** When the comet approaches the sun the nucleus or densest part never appears in the middle point as it certainly would do in obedience to gravity, if the interior arrangements of the particles were undisturbed by any force from without. Sometimes a vast mass of matter streams from the comet in a direction away from the sun; sometimes, as in Encke's comet, the body takes an oval shape with the nucleus near one edge. As the tail uniformly turns from the sun we cannot doubt that the sun is the seat of the disturbing force.

Does it then drive away the tail or attract the nucleus. Evidently it does neither only. If without compensation the sun repelled any portion of the mass of a comet, the gravitation of the whole comet toward the sun would be diminished. If the sun exercised any new or peculiar attraction on the nucleus, the gravitation of the whole mass would be increased. Either of these effects alone would produce a marked change in the comet's motion in its orbit. In the former case this would decrease, in the latter it would increase. Neither of these changes take place.

§ **162.** It occurred to Bessel that the sun might have both an attractive and a repulsive influence on the comet, one exactly balancing the other; and although deranging the comet's internal constitution, not affecting its gravitating tendency or its motions as a whole.

Many forces act in this way, producing not a single effect, but two opposite and compensating ones. Magnetism is of this kind. If the sun places the comet in a condition like that in which a magnet places a needle, its gravity would be undisturbed, but its mass would be endowed with polarity. The comet of Halley, during its last return, confirmed this view. As it approached the sun, very extraordinary activities appeared to affect its entire organization. At a very early period there was a singular outstreaming of light from the upper part of the nucleus towards the sun. This outstreaming mass soon showed itself to have a movement of oscillation or vibration exactly like that of a pendulum, causing it to swing from one side to the other of the line joining the nucleus with our luminary. This oscillation could be produced only by an attraction exercised by the sun; just as the swinging of the pendulum is owing to the attraction of the earth. It must either be a new power, or gravity attracting an irregular mass. The times of the oscillation being calculated, it was found that it could not be attributed to gravity. It is more probable that the sun cast the entire mass into a state of polarity. Much excitement was visible, and it seemed like a body being magnetized by induction. Bessel thought that when saturated, the luminous matter which had been thrown out toward the sun, turned round and enveloping the nucleus formed the tail.

§ **163.** The influence of the sun being doubtless invariable, the various effects produced by it on the shape of comets, points out varieties in their physical constitution. On one he bestows a slightly elongated form, on another a tail streaming through spaces wide as the earth's orbit; he causes a third to spread itself out like a fan.

The following account, given by Sir John Herschel, of Halley's comet, as he observed it at the Cape in 1835, gives a lively idea of the changes caused in its physical appearance by the sun. When first seen by him it appeared as a star of the third magnitude, hazy, and with a scarcely perceptible tail. The next night a crescent-shaped cap was formed on the side next the sun, the coma decidedly extending beyond it. These appearances continued, with some variations, till it passed its perihelion. Twenty-four days after its disappearance it was again seen, in its return from the sun, as a small star of the third magnitude, dim and hazy and with no tail. Viewed through the twenty-feet reflector, it was now a most surprising object. Its head was terminated sharply like the ground-glass shade of an argand lamp. Within the well-defined head, and somewhat eccentrically placed, was an object resembling a miniature comet, having a nucleus, head, and tail of its own, perfectly distinct, and considerably exceeding in intensity of light the nebulous head. At this time the comet was increasing in dimensions with such rapidity that it might almost be said to be seen to grow. Measurements of the diameter of the head taken within two hours and a quarter of each other differed sensibly. The next night its increase was evident at the first glance; its form had become elongated and less definite toward the tail; the coma had not increased proportionally and was much less bright. The whole bulk of the comet, exclusive of the coma, had considerably more than doubled within twenty-four hours.

§ **164.** By the 31st of January the coma had entirely disappeared; the interior comet was so much dilated as more than to fill the field of view (15°) in length, and nearly so in breadth; its outline was soft, rounded, well defined. From this night may be dated the commencement of the developement of the true tail, that is of the

10

prolongation into a regular train of the parabolic envelope, aided by the similar prolongation of the ray or internal comet. The coma from this time appeared no more; it was neither dissipated nor absorbed, but swept off by the sun's action into the tail. But in the progress of the comet towards extinction, the semblance of a new coma arose from the dilatation of the mass of internal light immediately surrounding the nucleus, which at last constituted the whole visible comet, the infinitely minute and hardly perceptible nucleus excepted.

The comet on the 31st was full of small stars; but their light was not extinguished by it. Innumerable small stars passed at various times extremely near to the nucleus, though none exactly on it, but were no more affected than they would have been by so much lamplight artificially introduced.

While the comet was measurable after its perihelion passage its dilatations were nearly uniform. Calculating backward, at the same rate, the envelope must, on the 21st of January, at 10 minutes P. M., have had no magnitude. Previous to that instant the comet must have consisted of a mere nucleus or stellar point more or less bright, and a coma more or less dense and extensive. At that instant the formation of this envelope and of the ray or internal comet commenced. The perihelion passage took place on the 15th of November, and it was not until the eighty-third day after that event that the formation of the envelope commenced. During these eighty-three days the comet must have been cooling, and must have arrived at the dew-point of the vaporous substance which composes the envelope.

§ **165.** We have now to consider the motions of comets. All the planets and all the satellites, so far as we know, revolve in orbits of one kind, in ellipses. Among comets there is probably a variety. Two small comets revolve in ellipses, and return regularly, and are considered as belonging to our system. Probably many others revolve in extremely elongated ellipses. Parabolas, hyperbolas, and exceedingly elongated ellipses, are so nearly alike at the part nearest the sun, that it is difficult to ascertain in which of these curves a comet moves. Successive observations, proving the reappearance of the same comet, may prove the elliptic form to prevail. Only two have been completely computed whose orbit is best represented by a hyperbola. The chance that comets describe parabolas is infinitely small compared to that of their describing ellipses or hyperbolas. To produce the former curve one particular velocity is necessary, the slightest increase or diminution of which will cause it to deviate into one or other of the two latter curves.

It may be doubted whether the comets of our system have always belonged to it, or whether they were visitants, and are detained by some causes unobserved by us. We may have an opportunity of settling whether this ever takes place by observation at some future time. At present we know nothing of their history, we know only that they now revolve about the sun in regular ellipses.

§ **166.** The periodic time of Halley's comet is between seventy-five and seventy-six years. Its orbit is a lengthened ellipse, extending far beyond that of the planet Uranus, and inclined to the ecliptic at an angle of between 17° and 18°. Its aphelion is about the distance of Venus from the sun. The next is the comet of Encke; its revolution is completed in 1,207 days, or 3⅓ years. It revolves in an ellipse of great eccentricity, at an angle of about 13° 22′ to the plane of the ecliptic. The third has a period of six years and eight months. The orbit of Biela's comet extends somewhat beyond Jupiter. At its perihelion, however, it approaches nearer to the sun than the earth. Encke's comet, at its perihelion, is about as distant from the sun as Mercury; at its aphelion, not quite so far as Jupiter. The number of known periodical comets is yearly increasing. Halley's comet, in its first recorded appearances, in 1305, 1456, &c., exhibited a brilliant tail. The apparent size and the length of the tail seem to have undergone diminution in its later returns. Encke's comet is not at all conspicuous, and Biela's is also small and without a tail, or any appearance of a solid nucleus whatever. The orbit of Biela very nearly intersects that of the earth, and had the latter, at the time of its passage in 1832, been a month in advance of its actual place, it would have passed through the comet.

§ **167.** These periodical comets are of particular interest, because they have almost and will hereafter quite settle a question much agitated among philosophers; whether the inter-planetary space is filled with an extremely subtle medium, or whether it is a vacuum. Some phenomena make either probable. If we receive the undulatory theory of light, as is now almost universally done, we require a medium stretching not only through the planets but to the remotest stars; for how otherwise can the undulations be propogated. On the other hand, if such a medium exists, we should suppose that the planets would have their motions altered by it in a degree which would in process of time become appreciable. But no period is so constant as the revolution of a planet. Each one accomplishes its revolution, and has done so for hundreds of years, in precisely the same time. A medium, however, which would impress no delay on the solid, weighty mass of a planet, may produce a very perceptible difference in the time of the revolution of a comet. Now it was found by Encke that the comet which bears his name had been constantly anticipating the calculated time of its arrival at its perihelion; in some instances two days, in others one day. Its ellipses are continually diminishing, its mean distance from the sun dwindling by slow but regular degrees. This is evidently the effect which would be produced by the resistance of a very rare ethereal medium pervading the regions through which comets move. For such resistance, by diminishing its actual velocity, would diminish its centrifugal force, and thus allow the sun power to draw it nearer. There is no other mode of accounting for the phenomenon in question, and accordingly it is the solution generally received. The comet will probably ultimately fall into the sun, should it not first be dissipated altogether, a calculation not at all improbable, considering the lightness of its materials.

Of comets not yet returned, we cannot know whether to consider them of our system or not. Perhaps they pass the long void which separates one fixed star from another, returning after ages to the same centre. Perhaps if the sun is translated in space, comets formerly visible may be left out of the sphere of his attraction, and the sun entering new groups or streams of comets may give them new orbits. It has even been supposed, by Arago, that a great number of comets might at a distance assume a nebulous appearance. What is the density of this ether and the law of its density near the sun, whether it is at rest or in motion, if the latter, in what direction it moves, are questions which comets must answer for us. If it revolves round the sun it must accelerate some comets and retard others; and by repeated observations on comets moving in different planes and diameters, the plane and rate of rotation may be ascertained. Halley's comet has been retarded in every successive return, and this explanation of its delay has been offered; but we know but little about it.

§ **168**. The periods of those comets which reappear more seldom can be determined with great difficulty, and as yet with no exactness. A period of 3,065 years has been assigned to the fine comet of 1811, and one of upwards of 8,000 years to the comet of 1680. If these periods are the true ones, these bodies recede to distances from the sun equal, one to twenty-one, and the other to forty-four times the distance of Uranus, or to 33,600 and 70,400 millions of miles. Even at these distances they feel the sun's attraction. But while the motion of the comet of 1680 at its perihelion is 212 miles a second, thirteen times that of the earth, its velocity at its aphelion is scarcely ten feet in a second, only three times that of the most sluggish rivers, and one half that of the Cassiguian, an arm of the Orinoco. It approaches 163 times as near to the sun as the earth does, and experiences a heat 26,000 times that of the earth; and since red-hot iron has only twelve times the heat received from the summer tropical sun, this comet is exposed to a heat 2,000 times as great as red-hot iron. This comet's distance from the sun is only forty-four times that of Uranus, while the nearest fixed star has 250 times that distance. Outside of this perhaps many comets revolve whose major axes are longer than that of the comet of 1680.

Having considered the greatest known distances of comets, we will now notice instances of their greatest proximity hitherto measured. This same comet of 1680 approached the sun's surface within one sixth of his diameter. Perihelia which take place beyond the orbit of Mars cannot often be observed from the earth, yet we have no reason to suppose that more lie within than without it. We have an opportunity therefore to observe very few of those comets which actually enter the solar system.

§ **169.** The mutual influence of comets and planets has always been a subject of great interest. By these the comet's progress may be retarded or accelerated, the place of its nodes changed, its perihelion distance diminished or increased, and the inclination as well as eccentricity of its orbit altered. And these changes during one revolution are sometimes so considerable as to render the identity of a comet at its successive returns to the sun very doubtful.

Halley's comet first drew the attention of astronomers to these perturbations. After having ascertained its approaches to the sun in the years 1531, 1607, and 1682, Halley was surprised to find that the period of its first revolution was longer by thirteen months than the following one. He thought the difference might be caused by the disturbing action of the planets, particularly Jupiter and Saturn; and after a rough estimate of their attractions he announced the return of the comet for the end of 1758 or the beginning of 1759. The comet appeared as announced, proving the weight of comets and the extent of gravitation to them. During its next revolution this comet will be very much diverted from its course by the planet Uranus.

§ **170.** The comet of 1770 exhibited still more remarkable changes in its orbit. Astronomers had in vain endeavored to represent its observed course by a parabola. At length its orbit was discovered to be an ellipse, not so elongated as to approximate to a parabola, but much shorter, and requiring only a period of five and a half years. This result seemed very extraordinary, since the comet which should to have been so often visible, on account of the shortness of its period and perihelion distance, had never yet been seen on any previous occasion; and the circumstance was still more unaccountable, when it was found that the comet made no subsequent return to the sun.

At length, by tracing back the movements of this comet in its orbit, it was found that at the beginning of 1767 it had entered within the sphere of Jupiter's attraction. The amount of this attraction being calculated from the known proximity of the two bodies, the previous orbit of the comet was determined. It must have been an ellipse of greater extent, having a period of fifty years, in which the comet, even when nearest the sun, was still as far distant as Jupiter. It was therefore very evident, why, as long as the comet continued to circulate in this orbit so far from the centre of the system, it never became visible from the earth; and also that the cause of its appearance in 1770 was the disturbing action of Jupiter which constrained it to move in a shorter ellipse and at a less distance from the sun.

§ **171**. Another question of interest is whether comets can act on planets so as to produce perturbations in their course; and also what would be the consequence of a collision between a comet and a planet. The comet of 1770 affords an answer to the first question. From its brilliancy this comet must have been of considerable size, and was even computed to have a diameter nearly thirteen times as large as the moon. On the two occasions above mentioned it is said to have traversed the whole system of Jupiter's satellites, and at each time required four months to free itself from the sphere of his attraction. Yet not the slightest alteration was observed in the motions of these small bodies. The same comet approached so near the earth as to shorten its own revolution by two days, yet what was its reaction on the earth? If its mass had been equal to that of the earth, it would have lengthened our year by two hours, forty-seven minutes. But nice calculations prove that in the length of that year no change exceeding two seconds would have taken place. Hence as 10,027″ : 2″ : : mass of earth : mass of comet. The comet's mass was less than the $\frac{1}{5000}$ part of the mass of the earth. It is evident therefore that none of the planets are liable to be carried out of their course by these diminutive bodies.

Other dangers have been apprehended from the approach of comets. It has been feared that the waters of the ocean would be attracted and thus form a deluge. The small mass of comets precludes all danger of this sort. Besides, the ocean would require some time before its inertia would be overcome; and meanwhile the rapid motion of the comet and the rotation of the earth would have presented a different surface of water to the comet.

§ **172.** But though proximity is not alarming, it is very different with actual contact. The risk of actual contact is infinitely small when we consider the immense extent of the planetary spaces. Still collision is possible, and its consequences not without interest to us. If the comet and planet were both moving toward the same quarter of the heavens each would glide from the surface of the other, without any very important change in their movements or their physical constitution. But should the directions of their respective courses be directly opposite, the consequences would be far more serious and permanent. The inconsiderable mass of the comet would be compensated by its prodigious momentum, and the planet might be impeded or altogether arrested in its orbit. If the momenta of the two bodies were equal, the progressive motion of both bodies would be destroyed and they would fall into the sun.

We may perhaps see in the heavens such a collision or the consequences of it. The comet of Encke approaches nearer to the planets than any other; it approaches to 360,000 miles distance of Mercury. This circumstance makes a collision between it and Mercury not improbable.

CHAPTER IX.

PHYSICAL ASTRONOMY.

Analogies observable among the Planets. Their general form and their Atmospheres Internal state of our Globe. Central Heat. Theories accounting for the external appearance of the Earth and Moon. Objections to the theory of Central Heat. Supposed differences of temperature in space. Laplace's Nebular Theory.

§ **173.** Let us throw together all we know of the physical state of our system and of the fixed stars, and see if any light is shed on the former state of the universe. We have ascertained that matter exists in the three states in which we find it at the surface of the earth. It is solid in the planets and moons, and undoubtedly also in the nuclei of the sun and stars. It is aeriform at the surface of the sun and stars, as is proved by the polarization of their light, and also in comets, and in the atmospheres of many of the planets, and in ether. It exists in a liquid form in several of the planets, as is proved by the clouds floating in their atmospheres. But in the moon no water is present, and no atmosphere has been detected.

The atmospheres of the planets differ from one another as to color and density. The sun has one or more atmospheres apparently beside the zodiacal light and ether which may in some way belong to him. Of the solid forms of matter we know very little. In Mars the red color of the soil may be distinctly seen; a variety of color is also perceptible on the moon's surface. The shooting stars have the same composition as the earth. On the whole, appearances favor the idea that the composition of the planets is identical or only slightly varied.

§ **174.** The general form we have seen is the same in all. All likewise depart slightly from this general form. Their surfaces are irregular. Mountain peaks are discernible in the moon, in the nearer planets, and perhaps in the sun. Gravity accounts for their general form; some upheaving force must have caused the departures from it.

The upheaving force is much weaker than gravity ; it only roughens the surface.

We call our globe solid and surrounded by a liquid and a gaseous envelope. It may also have a liquid interior. We know that it increases in density towards the centre. But of the state in which the materials of its crust exist below a small depth we know nothing. In mines and springs the heat of the earth increases about 1° Fahrenheit for 54.5 feet. If we suppose the increase to continue in an arithmetical ratio, a stratum of granite would be in a state of fusion at a depth of twenty-one geographical miles, or at between four and five times the elevation of the highest summit of the Himalaya. Chemical combinations and the neighborhood of volcanoes and the heat imbibed from them, may account for some of the internal heat, but not for so constant an increase and to so great a depth as has been observed in many parts of the solid land. Central heat accounts for it better, and it has therefore been a favorite hypothesis. Central heat also accounts for other appearances on the surface of the earth. Before we adopt it, however, we must see whether no other cause could produce the same appearances on the earth, and we must seek in the other planets, and particularly in our neighbor the moon, for evidence of the existence of central heat there. We must also consider whether from what we know of the formation of the earth the central portions would be likely to remain fluid.

§ **175.** Direct researches on our own globe teach us but little. Man's eyes are turned outwards. Accumulating all the facts which the best telescopes reveal, with regard to the distant stars which, strange to say, seem to be undergoing more changes than the humble members of our system, scrutinizing the planets and especially the moon, we may come to some definite conclusion respecting the interior of our globe, and the identity of the materials of the suns and planets. Nay we may form some idea of the circumstances under whose control they took their present shape, and may judge which are the older, which the newer, inhabitants of the heavens.

It is absolutely impossible to explain by central heat the present appearance of the moon's surface. In the moon the upheaving force is in a far greater ratio to gravity than in the earth, hence the mountains of the moon bear a much larger proportion to her size than those of the earth to the earth's size. If the upheaval was caused by the outbreak of a central fluid mass in one planet, it probably was in all others. A central fluid opening through cracks running nearly in great circles of the earth accounts very well for the Andes and the Himalaya, and other terrestrial chains, but accounts not at all for the lunar mountains, which are very differently disposed.

Let us consider whether local heat may not account more satisfactorily for the mountains both of the earth and the moon. Local heat may be generated in the crust of the earth, by water soaking through and coming in contact with matter in which it excites violent chemical action. The materials once combining under the immense pressure of the rocks above, all their gases being kept in, a heat would be caused sufficient to melt the hardest substances. A large lake of subterranean lava would be formed which would uplift plains, throw up mountains, and at length vent itself in a volcano.

§ **176.** Another theory supposes that the subterranean lakes of lava are remains of the fluid world. Without pretending to know what the interior of the earth and moon is we may suppose that for ten miles from the surface their crust is chiefly solid. We will allow the moon to be in the earliest stage of formation, as this suits better her present state. Suppose each orb contains in its solid crust liquid and gaseous patches, as some crystals contain water. In early stages the crust would bubble all over like yeast or dough. These bubbles breaking would form cavities like the moon's hollows. When all the more external and smaller bubbles of the crust have broken, periods of comparative rest would ensue. The earth may now be in this state. The melted masses that now lie deeper may sometimes be chemically disturbed and cause earthquakes and volcanoes. The earth has some traces of the state in which the moon is. The trap rocks, those which have been injected from below into the cracks of other rocks, appear to run from centres. In like manner from centres in the moon proceed rays which have been called lava, and supposed to be poured out from centres. This cannot be the case because they pass over hill and valley, appearing on the sides and at the bottom of precipices like dykes. Perhaps this theory explains as well as any other the great number of peaks and hollows in the moon and the circular form of the few chains discovered there.

§ **177.** The supposition that the interior of the earth is a fluid heated mass, agrees well with the nebular theory that the earth was once a heated mass of vapor. It is not however required by it. If the earth was originally fluid it might become solid by either of two modes, from cooling or from pressure. The heat would be continually dissipated from the surface, and would therefore be greatest at the centre ; and so long as the mass was fluid, the inequality of the heat would cause a constant circulation between the surface and the centre. Now, if the effect of heat in preventing solidification were greater than the effect of pressure in promoting it, solidification would begin at the surface, where a crust would be formed, and would constantly increase in thickness, by layer after layer added to its underside. But if the effect of pressure in promoting solidification were greater than the effect of heat in preventing it, solidification would begin at the centre and extend outwardly. While the process was going on, circulation would continue in the fluid part exterior to the solid nucleus. But before the last portions became solid, a state of imperfect fluidity would arise just sufficient to prevent circulation. The cooled particles at the surface being then no longer able to descend, a crust would be formed, from which the process of solidification would proceed far more rapidly downwards than upwards from the solid nucleus. Our globe would thus arrive at a state in which it would be composed of a solid exterior shell, and a solid central nucleus, with matter in a state of fusion between them.

§ **178.** If the earth when first thrown off, or when it first began to take form, had been suddenly transported into regions very much colder than those it left, the outer crust might have been solidified rapidly, and have thus imprisoned the melted mass. But from the arrangement of the solar system it appears improbable that any such violent change of temperature took place.

But if the solidification was owing to pressure as much or more than to change of temperature, it would begin at the centre and extend gradually to the surface. The pressure of gravity would begin to act as soon as the mass was insulated. It would vary from nothing at the surface to a pressure probably surpassing 100,000 the pressure of our present atmosphere. This pressure would reduce all the layers of vapor to the solid state, beginning with the central masses, and proceeding toward the surface till nothing remained unsolidified but our sea and air.

§ **179.** This change would not be instantaneous, for time would be required for each bed to be pushed toward the centre. Radiation of heat is so extremely rapid that the beds or concentric strata of the earth would get rid of the heat developed during their solidification, by radiating it through the upper beds still in a vaporous form. So that there need not be supposed to remain now, or at any definite point of past time, the slightest trace of that heat, how great soever it may have been. An effect similar to that now described would ensue, in the case of a cylinder of great length closed at both ends and filled with the vapor of water at the maximum density corresponding to the external temperature. Were the cylinder horizontal, the weight of the fluid could have no influence ; but if it were raised up and placed vertically on one of its ends, its weight would produce a pressure on its different beds, increasing from the top downwards ; and through effect of this pressure, liquefaction would take place, beginning on the lowest part of the tube and proceeding upwards. The time occupied by each bed or stratum in descending would not be easy to determine ; but certainly it would suffice to permit the latent heat, developed by the bed liquified immediately before, to escape by radiating ; and thus the water evolved from the vapor would not be heated, but would simply preserve the temperature of the external space.

§ **180.** There is one great objection to the theory of central heat. We cannot but suppose that some of it would be all the time escaping. If so, in the course of ages the earth would become smaller. The velocity of the earth's rotation depends on her volume. Since, therefore, by the gradual cooling of the mass by radiation, the axis of rotation would become shorter, such decrease of temperature would be accompanied by increased velocity of rotation, and diminished length of day. Now the comparison of the secular inequalities in the moon's motion with eclipses observed by the ancients, shows that during an interval of two thousand years, the length of the day has not been diminished by the one hundredth part of a second. We know therefore that the mean temperature of the earth has not altered during that period so much as the $\frac{1}{306}$th part of a degree of Fahrenheit.

§ **181.** Abandoning therefore central heat as the cause of the elevated temperature of deep pits, two other explanations of it have been offered, founded on causes whose existence is probable, and which are capable of producing the observed effect. The heat of the different regions traversed by the earth, while moving with the sun and his system through space is unequal. The temperature of any part of space may be modified by others having specific temperatures, or it may be owing only to the radiant heat emanating from the different stars, and crossing it in all directions. How much the temperature of different parts of space varies, from being more or less thickly studded with stars, or how much the stars themselves may vary as to their heat-giving power, we know not. But we may fairly suppose that differences must exist, sufficient to account for the evidences of change of temperature we find in the earth. Through the extent of the earth's annual orbit the temperature of space shows no change. In order to perceive it, points at vast distances from one another must be compared. But the solar system occupies vast intervals of time in passing from one of these points to another. The changes from one region to the other are probably gradual. We could not therefore in a few years or perhaps centuries expect to perceive any decided change in our general climate. But if we found that the earth retained beneath its surface a store of heat, we should conclude that in passing from a hot to a cold region, on account of its massiveness, it had retained a portion of heat.

§ **182.** Now the earth presents indications of having been both warmer and colder than it now is. Fossil plants and animals are now found in high latitudes where living ones of similar character could not exist. To account for this we must have passed through a hotter region than the present. It must be acknowledged however that the hypothesis of central heat and of the gradual cooling of the earth through millions of years accounts equally well for this. But we have also indications that large portions of the northern hemisphere, (the same hemisphere in which fossil elephants and tropical plants are found,) were once covered with glaciers. The increase of cold which is required to account for the extensive formation of glaciers is only 14°, a quantity which might perhaps be accounted for by our passing from the neighborhood of an orb like Sirius to a desolate region. Yet after all our speculations it must be acknowledged that of no subject are we more ignorant than about the globe on which we live. Man's eyes are turned outward. Perhaps we are to learn ourselves from others. Perhaps the moon or some sister sphere will give us that knowledge of the interior of our globe which we vainly seek by penetrating its surface. The solution of the problem may be withheld from man while he lives on it, or the truth may be gradually evolved by the laying aside of error. Meanwhile both the astronomer and geologist must contribute to its further elucidation.

§ **183.** A theory was proposed by Laplace to account for the successive changes in the heavens, and to open the page of history from the creation of the universe.

His theory supposed a nebulous fluid to float in space, and from this suns and worlds to be gradually evolved and thrown off. It rested chiefly on the presence of large nebulous masses which were supposed to be slowly formed into suns and planets.

Lord Rosse's telescopes have resolved most of the nebulæ into stars, and thus rendered doubtful the existence of nebulous matter. But besides that the existence of the nebulosity, on which this splendid hypothesis has been reared, may reasonably be doubted, the difficulties which present themselves in comparing it with the actual state of the solar system, are numerous and grave enough to warrant the assertion that the name of the illustrious proposer of the nebular theory is after all its chief recommendation. So greatly, however, has it influenced the modern view of the formation of the solar and sidereal systems, that some knowledge of its main features is indispensable to a right understanding of much that has been written on that subject. The following exposition of Laplace's Cosmogony, given by Pontecoulant, is therefore here introduced.

§ **184.** It seems a great deal to have discovered the true laws of the celestial motions, and to have been able to assign, with so much probability that it almost amounts to certainty, the cause which produces them. It might have been thought that man ought to rest here; but in science, as in every other thing, success more frequently rouses ambition than satisfies it. Master of the great secret employed in nature to give life to the planetary system, this glory has not sufficed him; he has sought, by going back to past ages, to pierce even the mystery of its formation, and he has dared to conceive the bold thought of being present, if we may say so, at the spectacle of the creation of the world.

Buffon was the first to start this vast question, and to consider from this philosophical point of view the constitution of the universe. His ideas on the primitive formation of the planets and satellites would find few supporters among astronomers now. He supposes that the force of a comet falling obliquely on the sun, has projected to a distance a torrent of the matter of which it is composed, as a stone thrown into a basin causes the water which it contains to gush out. This torrent of matter, in a state of fusion, has broken into several parts, which have been arrested at different distances from the sun, according to their density or the impetus they received; they then united in spheres, by the effect of motion of rotation, and, condensing by cold, have become opaque and solid, and formed planets and satellites.

§ **185.** This system explains very simply the unity of direction of the orbitual motion of the planets, which all circulate from west to east, round the sun; but it is not the same with the rotary motion, and we can find no reason why this motion should be in the direction of the orbitual motion rather than in the contrary one.

This identity of the direction of the rotary and orbitual motions of the planets and satellites, is one of the most remarkable facts pointed out by observation. We see no better why the orbits of the planets are all nearly circular and comprised in a narrow zone of the celestial sphere, whilst the comets move in orbits very eccentric, and with any inclination whatever to the ecliptic. The hypothesis of Buffon is thus very far from explaining the principal phenomena which characterize the planetary system, and cannot now merit a serious examination.

§ **186.** Let us try, with Laplace, to find out their true cause in another way. This new system has for its support the labors of Sir W. Herschel, aided by his powerful telescopes, in regard to the *nebulæ*, by which appellation those whitish spots, seen in different parts of the heavens, of which they occupy a large extent, have been called. Observing these spots attentively, the nebulous matter is at first seen in a most diffused state, and reflecting only a feeble and almost uniform light; in others this matter is condensed round one or several dim parts; in others these centres are more brilliant, in proportion to the nebulous matter surrounding them; afterwards the atmosphere of each body separating, by an ulterior condensation, there result numerous nebulæ, formed of brilliant bodies near each other, and each surrounded by an atmosphere; at last, a higher degree of condensation changes these nebulæ into stars. Classing together these observations on a great number of different nebulæ, Herschel supposes that they represent a series of operations on a single mass of nebulous matter, which would pass from its first state of completely diffused, scarcely luminous nebulosity, to the state of the most brilliant stars. The progress of the condensation effected by this change, could only become perceptible after the lapse of centuries; but we may discover it by examining at once the whole of the nebulæ diffused through the sky, as a naturalist, who wishes to discover the successive developements of the organs of an animated being, studies them in individuals of different ages.

Since the attentive observation of the nebulæ seems to show their change into stars, at epochs more or less remote, we must suppose from analogy that the existing stars and the sun himself were formerly masses of nebulous matter, reduced by condensation to the state in which we now see them. From this induction we are led to regard the sun, at the origin of things, as composed of a body more or less brilliant, surrounded by a vast atmosphere, which extended at first, by the effect of excessive heat, beyond the orbits of all the planets, and was confined successively by condensation to its actual limits.

§ **187.** The atmosphere, which we may suppose possessed of a rotary motion round its centre of gravity, whether this motion results from the reciprocal attraction of all its parts, or has been communicated to it primarily, must, in condensing by cold, leave in the plane of its equator zones of vapor composed of substances which required an intense degree of cold to return to a liquid or solid state. These zones must have begun by circulating round the sun in the form of concentric rings, the most volatile molecules of which have formed the superior part, and the most condensed the inferior part. If all the nebulous molecules of which these rings are composed, had continued to cool without disuniting, they would have ended by forming a liquid or solid ring. But the regular constitution which all parts of the ring would require for that, and which they would have needed to preserve whilst cooling, would make this phenomenon extremely rare. Accordingly the solar system presents but one instance of this, that of the rings of Saturn. Generally the ring must have broken into several parts, which have continued to circulate round the sun, and with almost equal velocity, whilst at the same time, in consequence of their separation, they would acquire a rotary motion round their respective centre of gravity; and as the molecules of the superior part of the ring, that is to say, those farthest from the centre of the sun, had necessarily an absolute velocity greater than the molecules of the inferior part which is nearest it, the rotary motion, common to all the fragments, must always have been in the same direction as the orbitual motion.

§ **188.** However, if after their division one of these fragments has been sufficiently superior to the others to unite them to it by its attraction, they will have formed only a mass of vapor, which, by the continual friction of all its parts, must have assumed the form of a spheroid flattened at the poles and elongated in the direction of its equator. Here then are rings of vapor left by the successive retreats of the atmosphere of the sun, changed into so many planets in the condition of vapor circulating round the sun, and possessing a rotary motion in the direction of their revolution. This must have been the most common case; but that in which the fragments of some ring would form several distinct planets possessing different degrees of velocity, must also have taken place, and the four telescopic planets, Ceres, Juno, Pallas and Vesta, discovered at the beginning of the present century, seem to present an instance of this; at least, if it is not admitted, with Olbers, that they are the fragments of a single planet, broken by a strong interior commotion. It is easy to imagine the successive changes produced by cooling on the planets whose formation has just been pointed out. Indeed, each of these planets, in the condition of vapor, is, in every respect, like one of the nebulæ in the first stage; each must, therefore, before arriving at a state of solidity, pass through all the stages of change we have just traced in the sun.

§ **189.** At first, the condensation of its atmosphere will form round the centre of the planet a body composed of layers of unequal density, the densest matter having, by its weight, approached the centre, and the most volatile reached the surface, as we see in a vessel different liquids ranged one above another, according to their specific gravity, arrive at a state of equilibrium. The atmosphere of each planet will, like that of the sun, leave behind it zones of vapor, which will form one or several secondary planets, circulating round the principal planet, as the moon does round the earth, and the satellites round Jupiter, Saturn and Uranus, or else they will form, by cooling without dividing, a solid and continuous circle, of which we have an instance in the rings of Saturn. In every case, the direction of the rotary and orbitual motion of the satellites or the ring, will be the same as that of the rotatory motion of the planet; and this is completely confirmed by observation.

§ **190.** The wonderful coincidence of all the planetary motions, a phenomenon which we cannot, without infringing the laws of probability, regard as merely the effect of chance, must then be the result of the formation of the solar system in this ingenious hypothesis; we see also why the orbits of the planets and satellites are so little eccentric, and deviate so little from the plane of the solar equator. A perfect harmony between the density and temperature of their molecules in a state of vapor, would have rendered the orbits rigorously circular and made them coincide with the plane of this equator; but this regularity could not exist in all parts of such large masses; there have resulted the slight eccentricities of the orbits of the planets and satellites, and their deviation from the plane of the solar equator.

When in the zones, abandoned by the solar atmosphere, there are found molecules too volatile either to unite with each other or with the planets, they must continue to revolve round the sun without offering any sensible resistance to the motions of the planetary bodies, either on account of their extreme rarity, or because their motion is affected in the same way as that of the bodies they encounter. These wandering molecules must thus present all the appearances of the zodiacal light.

§ **191.** We have seen that the figure of the heavenly bodies was the necessary result of their fluidity at the beginning of time. The singular phenomenon presented by the rigorous equality indicated by observation, among the lesser motions of rotation and revolution of each satellite, an equality rendering the opposed hemisphere of the moon forever invisible to us, is another obvious consequence of this hypothesis. Indeed, supposing that the slightest difference had existed between the mean motion of rotation and revolution of our satellite whilst it was in the state of vapor or of fluidity, the attraction of the earth would have elongated the lunar spheroid in the direction of its axis towards the earth. The same attraction would have tended to diminish insensibly the difference between the rotary and orbitual motions of the moon, so as to confine it to narrow limits.

§ **192.** The principal phenomena of the planetary system are therefore explained with great facility by the hypothesis we are examining; and as these successive changes of a nebulous mass, and the leaving of a part of its substance by cooling, agree with all the leading phenomena, it must be allowed a high degree of probability. In this hypothesis the formation of the planets would not have been simultaneous; they have been created successively at intervals of ages; the oldest are those which are farthest from the sun, and the satellites are of a more recent date than their respective planets. It may be, if we are ever permitted to reach so high, that by an examination of the constitution of each planet, we may go back to the epoch of its formation, and assign to each its place in the chronology of the universe. It is likewise seen that the velocity of the orbitual motion of each planet as it is now, must differ little from that of the rotary motion of the sun, at the period when the planet was detached from its atmosphere. And as the rotary motion is accelerated in proportion as the solar molecules are confined by cooling, so that the sum of the areas which they describe round the centre of gravity would remain always the same, it follows that revolutionary motion must be so much more rapid as the planet is nearer the sun; and this is seen by observation.

§ **193.** It likewise results that the duration of the rotation either of the sun or of a planet, must be shorter than the duration of revolution of the nearest body which circulates round them; this observation is completely confirmed even in those cases where the difference between the duration of the two motions must be very slight. Thus the interior ring of Saturn being very close to the planet, the duration of its rotation must be almost equal, but a little longer than that of the planet. The observations of Herschel give indeed 0.432 of a day as the duration of the rotation of the ring, and 0.427 of a day as that of the planet; why then should we not admit that this ring has been formed by the condensation of the atmosphere of Saturn, which formerly extended to it? We may perhaps deduce from the laws of mechanics, and the actual dimensions of the sun, and the known duration of its rotation, the relation existing between the radius vector of its surface and the time of its rotation in the different stages of condensation through which it has passed. The third law of Kepler would be no longer the mere result of observation; it would be directly deduced from the primordial laws of the heavenly bodies.

§ **194.** In this system, as in that of Buffon, the particular form of the planets, the flattening at the poles, and bulging out at the equator, is only the necessary consequence of the laws of the equilibrium of fluids, and easily explains the greater part of the phenomena observed by geologists in the constitution of the terrestrial globe, which appear inexplicable, if it is not admitted that the earth and planets have been originally fluid.

Let us now see what is the origin and part assigned to comets by this hypothesis. Laplace supposes that they do not belong to the planetary system, and he regards them as masses of vapor formed by the agglomeration of the luminous matter diffused in all parts of the universe, and wandering by chance in the various solar systems. Comets would thus be, in relation to the planetary system, what the aerolites are in relation to the earth, with which they seem to have no original connection. When a comet approaches sufficiently near the regions of space occupied by our system, to enter into the sphere of the sun's influence, the attraction of that luminary, combined with the velocity acquired by the comet, causes it to describe an elliptic or hyperbolic orbit. But as the direction of this velocity is quite arbitrary, comets must move in every direction and in every part of the sky.

§ **195.** The cometary orbits will then have every inclination to the ecliptic; and this hypothesis explains equally well the great eccentricity by which they are usually affected. Indeed, if the curves described by comets are ellipses, they must be greatly elongated, since their major axes are at least equal to the radius of the sphere of the sun's attraction; and we must consequently be able to see only those whose eccentricity is very great and perihelion distance inconsiderable; all others, on account of their minuteness and distance, must always be invisible; unless, at least, the resistance of the ether, the attraction of the planets, or other unknown causes, diminish their perihelion distance, and bring them nearer the terrestrial orbit. The same circumstances may change the primitive orbits of some comets into ellipses whose major axes are comparatively small; and this has probably happened to the periodical comets of 1759, 1819 and 1832. The laws of curvilinear motion likewise show that the eccentricity of the orbit chiefly depends on the direction of the comet's motion on its entering the sphere of the sun's attraction; and as this motion is possible in every direction, there are no limits to the eccentricities of the orbits of comets.

If, at the formation of the planets, some comets penetrated the atmospheres of the sun and planets, the resistance they met would gradually destroy their velocity; they would fall on those bodies describing spirals, and their fall would have the effect of causing the planes of the orbits and equators of the planets to remove from the plane of the solar equator. It is, therefore, partly to this cause, and partly to those we have developed above, that the slight deviations we now perceive must be attributed.

§ **196.** Such is the summary of the theory of Laplace, on the origin of the solar system. This hypothesis explains, in the most satisfactory manner, the three most remarkable phenomena presented by the planetary motions.

1st. The motion of the planets in the same direction, and nearly in the same plane.

2d. The motion of the satellites in the same direction as their planets.

3d. The singular coincidence in direction of the rotary and orbitual motions of the planets and the sun, which in other systems would present inexplicable difficulties.

The no less remarkable phenomena of the smallness of the eccentricities and inclinations of the planetary orbits are also a necessary consequence of it, whilst we see, at the same time why the orbits of the comets depart from this general law, and may be very eccentric, and have any inclination whatever to the ecliptic. The flattening of the form of the planets, shown on the earth by the enlargement of degrees of the meridian, and by the regular increase of weight in going from the equator to the poles, is only the result of the attraction of their molecules whilst they were yet in a state of vapor, combined with the centrifugal force produced by the rotary motion impressed on the fluid mass. In short, among the phenomena presented by the motions and the form of the heavenly bodies, there are none which cannot be explained with extreme facility by the successive condensation of the solar system ; and the more this system is examined, the more we are led to acknowledge its probability.

§ **197.** Undoubtedly, if, as Laplace has himself said, a hypothesis not founded on observation or calculation must always be presented with extreme diffidence, this, it will be granted, acquires, by the union and agreement of so many different facts, all the marks of probability. But what principally distinguishes it from the ordinary theories concerning the formation of systems, is the identity which it establishes between the solar system and the stars spread so profusely through the sky.

All the phenomena of nature are connected, all flow from a few simple and general laws, and the task of the man of genius consists in discovering those secret connections, those unknown relations, which connect the phenomena which appear to the vulgar to have no analogy. In going from a phenomenon of which the primitive law is easily perceived, to another in which particular circumstances complicate it so as to conceal it from us, he sees them all flowing from the same source, and the secret of nature becomes his profession. Thus the laws of the elliptic motion of the planets led Newton to the great principle of universal gravitation, which he would have sought for in vain in the less simple phenomena of the rotary motion of the earth, or the flux and reflux of the sea.

§ **198.** But, this great principle being once discovered, all the circumstances of the planetary motions were explained, even in their minutest details, and the stability of the solar system was itself only the necessary consequence of its conformation, without which, as Newton thought, God would be constantly obliged to retouch his work, in order to render it secure. Laplace, extending to all the stars, and consequently to the sun, the mode of condensation by which the nebulæ are changed into stars, has connected the origin of the planetary system with the primordial laws of motion, without recurring to any hypothesis but that of attraction. He has, therefore, extended the great law of universal gravitation, which is probably the only efficient principle of the creation of the physical world, as it is of its preservation.

The hypothesis of Buffon required not only the fall on the sun of a comet as large as the mass of all the planets and satellites, which is very improbable, but in order to explain the formation of the innumerable planetary systems which the imagination may conceive round each star, as round the sun, it would have been necessary to imagine the fall of so many new comets, that reason would soon refuse to believe in chances so often repeated, and always when there was need of them. The principle of the condensation of the atmosphere of the nebulæ is, on the contrary, general, and would produce phenomena nearly analogous, in all the stars and planets.

CHAPTER X.

THE ROTATION AND FIGURE OF THE EARTH.

Effect of the Motion of the Earth on the apparent Motion of the Heavens. Distinction between the Earth's yearly and diurnal Motion. Axis of Rotation. The Circles of the Equator and Ecliptic. Shape of the Earth's Orbit. Invariability of the Earth's Rotation. Earth's Figure and Dimensions. Its Ellipticity. Pendulum experiments. Measurement of Arcs of a Degree on the Earth's surface.

§ **199.** The reader will remember those groups of islands which stud the Pacific. On the map they are near neighbors, but in fact they are too far apart to be visible to one another, and are separated by thousands of miles of ocean. Thus the inhabitants of each group might easily suppose their own to be the only group of the Pacific. And as the islands of each group are likewise widely separated from one another, it would not be strange if each island should be deemed by its inhabitants a world in itself.

Let us choose from a group containing some hundreds of islands, an island not precisely in the centre of the group. And let us suppose it endowed with a magnetic power like the Black Mountain of fairy tales. We will suppose several boats of different sizes to circulate round it at distances varying from 3 to 190 yards. Let each boat also twirl continually round an imaginary axis, while it proceeds in its course round the island. Place upon this boat a tiny insect, whose life endures only some seventy revolutions round the island, and this insect, supposing it endowed with reason, will have greater facilities for learning the wide Pacific with all its isles than man has for becoming acquainted with the universe. The earth is our boat, which measures from end to end 8,000 miles or upwards of forty-two millions of feet. The island round which we circle is our sun, and it is so bright, that from the part of the boat toward him we can discern no other body in the universe. But when our side of the boat is turned from him we see other boats, the planets our companions, and we also see the myriad isles of our cluster, beaming fixed stars in the heavens.

§ **200.** Since the difficulties in the way of learning are so great we will begin modestly and study our own planet first. There are three questions which must be settled before we can take the earth as our stand-point, and our measure for other bodies. The first is whether it is at rest or in motion; the second, what is its shape; the third, what is its size?

On looking into the heavens we find nothing apparently at rest. The sun is not in the same place if we look at him after an interval of an hour; the moon, if she is visible in the heavens, changes her place with rapidity. The stars which at twilight we saw in the east are over our head at midnight, and at dawn set in the west. Not only this, but the sun no two consecutive days rises or sets in the same place; and a different set of stars make their appearance in the east each night. Venus appears sometimes as an evening, sometimes as a morning star. The other planets are often absent from our heavens.

§ **201.** The first division we should attempt to make of these motions is by finding which are common to all, and which are peculiar to some. Those which are common to all may arise from either of two causes, and we must choose the more probable of the two. Either all the objects we look at are moving or we may ourselves be in motion. An observer in a vessel on a wide ocean cannot distinguish whether another vessel within sight is sailing and his own at rest, or whether the other vessel is at rest and his own drifted along by some unsuspected current. An observer on one ship cannot judge accurately the rate of progress or the direction of the other. For supposing them both in parallel directions, the swiftest will appear to advance with only the difference of their velocities. Suppose them sailing in opposite directions, the other will recede with the sum of their velocities. If they are sailing in diverging or converging directions from one another, the apparent direction will not for one moment coincide with the true one.

But instead of being on the ocean let the observer be on a lake whose banks are covered with villages and varied by woods and hills. If he finds that all these objects pass in the same direction and at an equal rate from his sight, and that in a certain time he has seen them on every side of him, will he not more rationally ascribe the motion to his single vessel than to the whole landscape around him. And if he sees on the lake other ships, not joining in this motion of the shores, but outstripping or lagging behind, he will not hesitate to ascribe to these ships proper motions; remembering always that he does not see these motions as they really are, but altered by his own motion.

§ **202.** Thus when we are obliged to choose between the simultaneous motions of many bodies, or the motion of one body, we keep on the side of probability when we attribute the motion to the single body.

Applying this principle to our earth, it is much more probable that by turning round she is a part of the twenty-four hours exposed to the sun, and another part in the presence of the stars, than that the sun and all the stars sweep round her in one uniform diurnal motion. It is much more likely that she is so inclined in her orbit that in one season of the year she presents a given part of her surface but a short time, and in another season a longer time to the sun, than that he, varying his motion, describes now lower and shorter, and again higher and longer arcs in the heavens.

We are therefore warranted in considering the earth's real motion as the cause of all the apparent motions which are common to other bodies. And we may ascribe to the same cause that part of each body's apparent motion for which it will account. Thus we find in the heavens three kinds of motion, one purely apparent caused by the earth's real motion, one made up of the body's proper motion combined with the earth's motion, and one among the fixed stars too distant to be affected by the earth's motion, and for all we know a simple proper motion.

§ **203.** Let us now consider what motions the earth must have to account for all the appearances we observe. There are two sets of phenomena to be accounted for, the diurnal and the yearly; we must therefore allow her two motions. We find that the sun appears to pass through an entire circuit of the heavens in the course of the year, moving from west to east. This apparent path of the sun may be caused by a real motion of the earth in the same direction from west to east. For supposing the sun to be motionless and the earth to be moving round him, the inhabitants of the earth would refer the sun to a point in the concave surface of the sphere precisely 180° from the earth's place. If the earth moved 15° eastward, the sun would move 15° eastward; if in a year the earth passed through the whole circle, the sun would appear to pass through the whole circle, always remaining precisely in opposite quarters of the heavens to those in which the earth would be. All these phenomena would be precisely the same and would occur in the same order whether the sun moves round us or we move round the sun.

We have now to account for the daily rising and setting of the sun and stars. To do this we need only suppose in the earth rotation in the same direction with its revolution. As the earth rotating toward the east turns up a portion of her surface toward the sun, the sun appears to that portion to rise in the east. As the place comes under the direct beams of the sun it is noon at the place; as it turns further round, it leaves the sun behind, he becomes invisible, but the stars rise; the earth still turns round, the place is 180° away from the sun, it is midnight; the earth turns another quarter, the place has left the stars behind, they have set; it has come in sight of the sun, the sun has again risen. This motion of rotation is confirmed even by phenomena on the surface of the earth. A stone let fall from the top of a high tower, falls not at the foot of the tower, but a little further east, showing that being further from the earth's centre it had a greater velocity than the earth it fell on.

§ **204.** But while the earth has been turning it has passed on in its orbit, it refers the sun to a different part of the concave sphere, the sun rises among stars lying farther east than those among which it rose yesterday. The stars which rose after sunset the night before are now lost in his beams, those which rose later now appear as soon as he is out of sight. Those which were on the meridian at midnight now pass it before midnight, and new stars come into view before the sunrise.

The daily and yearly changes may be exhibited in miniature by a bright light placed in the centre of a room whose walls are covered with tapers. Let the light represent the sun, the tapers the fixed stars. By walking round this room and rotating once every minute we shall face the lamp and thus make it rise to us once every minute; we shall also have the tapers facing us a half of every minute. If we walk round the lamp so slowly that we can turn round 365 times before we return to our starting point, we shall perform a course similar to that of the earth during a year. For by our revolution round the lamp we shall have brought it in succession between ourselves and every part of the wall, while by our rotation we shall have brought the light facing us, that is we shall have had 365 noons, and 365 times we shall have had a partially different set of tapers exhibited to us in each rotation.

§ **205.** If the earth rotates it must have an axis or line of immovable particles passing through its centre; and the diameter of this axis, and the fact of its variability or invariability we can only ascertain from the heavens. An axis may be variable in its position within the sphere that is as to the place where its poles touch the surface, it may be invariable with regard to the solid sphere, and may vary in direction carrying along the sphere with it, or it may be invariable in both respects. In the latter case it will always point toward one point, and all surrounding objects will appear to be carried in circles round that one point. As we find one point in the northern celestial hemisphere and another in the southern, round which all the stars appear to describe circles, we have no hesitation in saying that the axis of our earth is immovable, and if prolonged would pass through these points. On the earth's surface and in a line with those two immovable points in the heavens are two points which do not rotate, and partake only of the orbitual motion of the earth. These are the north and south poles of the earth.

The equator is an imaginary great circle on the earth's surface 90° from each pole, dividing the earth's surface into two hemispheres. Its plane is perpendicular to the earth's axis.

The latitude of a place on the earth is its distance from the equator north or south. The polar distance or angular distance from the nearest pole is the complement of the latitude.

The intersection of the plane of the equator with the sphere of the heavens is called the equinoctial. The distance of a star north or south of the equinoctial is called its declination.

§ **206.** The secondaries to the equator are called meridians because that secondary which passes through the zenith of any place is called the meridian of that place, and is at right angles both to the equator and the horizon, passing as it does through the poles of both.

In the heavens these secondaries are called hour circles because the arcs of the equinoctial intercepted between them are used as measures of time.

Longitude on the earth's surface is reckoned from some arbitrary point. The English and Americans reckon from the observatory at Greenwich, 180° westward, and 180° east longitude. It is more convenient to reckon 360° westward.

As declination corresponds to latitude so does right ascension in the heavens correspond to terrestrial longitude. It is likewise reckoned from a point arbitrarily chosen, in the vernal equinox, where the equator cuts the earth's path in spring, the beginning of our northern year. If the star is situated in the equator its right ascension is the number of degrees of the equator between the star and the vernal equinox. But if the star is north or south of the equator, then its right ascension is the arc of the equator intercepted between the vernal equinox and that secondary to the equator which passes through the star.

§ **207.** The sun's apparent and the earth's real path as marked out among the stars is an (almost) invariable great circle, and is called the ecliptic. One half of it is north and the other half south of the equator. Its plane is inclined 23° 28′ to that of the equinoctial, consequently its poles lie 23° 28′ distant from those of the equinoctial.

Celestial latitude and longitude are to the ecliptic what declination and right ascension are to the equinoctial; and both longitude and right ascension are reckoned from the same point, the vernal equinox.

The latitude and longitude of the heavenly bodies are not observed directly on account of the difficulty of verifying the proper instruments intended for this purpose. When required they are deduced by calculation from the observed declination and right ascension. All the apparent motions caused by the earth's rotation are referred to the equinoctial and its secondaries immediately. All apparent motions arising from the earth's revolution are more conveniently referred to the ecliptic.

§ **208**. The axis of the earth is always parallel to itself and always apparently points near the polar star. For the diameter of the earth's orbit is as nothing in comparison with the distance of the fixed stars. Two parallel bars a few yards asunder both apparently point to the moon when in the horizon. And these two or three yards may bear to 240,000 miles, the moon's distance, a greater ratio, than 190,000,000 of miles do to our distance from the polar star. Thus in all parts of its orbit the axis of the earth points in the same direction, and from the immense distance of the star to which we refer it, compared with the small size of the earth's orbit, it appears to point to the same place. Thus the pole of the heavens is the vanishing point not only of all great circles perpendicular to the equator in one position of the globe but at all seasons of the year. North to a person at the equator is in a line perpendicular to the equator; but all these lines referred to the distant concave sphere unite and we call north a point. In the same way all the hour circles which could be drawn at all seasons of the year vanish in two opposite points of the equinoctial.

As all planes perpendicular to the equator vanish in the poles, so do all planes parallel to the equator vanish in the equinoctial; for the polar diameter of the earth can cause displacement in the heavens no more than the equatorial diameter. An observer at each pole and one at the equator will therefore see the equinoctial in the same place.

All planes parallel to the ecliptic vanish in the ecliptic; all planes perpendicular to it vanish in the poles of the ecliptic.

Thus the motions of the earth furnish us with two sets of circles by means of which we can determine the position of the heavenly bodies. We learn the ecliptic directly, and infer the place of its poles. We observe the position of the poles of the earth, and infer the place of the equator from them.

§ **209.** We must not however suppose that because the line which the earth describes in the sphere in a year is a circle, that its orbit is itself precisely circular. Its true form may be ascertained by measuring the sun's apparent diameter; or by observing his angular velocity at different parts of the year, and ascertaining whether this is such as would take place in a circular or in an elliptic orbit.

If the diameter of the sun be accurately measured it is found to be greater at some periods of the year than at others. If such observations be continued for several successive years, it will still further appear, that, at the same time in each year, his diameter will be equal to what it was the year before. We must then conclude, either that the sun regularly expands and contracts, or that our distance from him is variable. The former supposition is absurd. We must therefore adopt the latter. If the earth remained at the same distance from the sun at all periods of the year, that is, were its orbit circular, the sun's diameter would never vary. Since the sun's diameter does vary we are persuaded that its orbit is not a circle. And since the angle under which an object is viewed, varies inversely as the distance, the variation in the sun's diameter will enable us to determine what the form of the orbit is. If at the second observation, the diameter is half what it was at the first, the second distance will be double the first; if one third, the distance will be treble; and so on. If we begin on a particular day to measure the sun's diameter, and for every week for a whole year continue to do the same, and if we set off on paper lines radiating from a common centre, proportioned to these different measurements, by joining the extremities of those lines, we shall accurately represent the form of the earth's orbit.

§ **210.** For instance, draw a straight line and take in it a point, Plate I. Fig. 4, for the position of the sun. On the 21st of December the sun's diameter is $32\frac{1}{2}'$ its maximum; on the 21st of June it is about $31\frac{1}{2}'$ its minimum. Since the distance is inversely as the apparent magnitude, we will set off towards I, that part of the orbit where the earth is in our winter, from a convenient scale of 63 equal parts, (that being the number of half minutes in $31\frac{1}{2}'$.) We will also set off from the same scale 65 equal parts, (65 being the number of half minutes in $32\frac{1}{2}'$,) measuring them on the same straight line running through S, but on the part of it between S and the part of the earth's orbit occupied by it in summer. When the year has advanced six weeks, or one eighth of the whole, let the sun's diameter measure $32\frac{2}{5}'$. This indicates the proportional distance of the earth six weeks before or after mid summer. Draw a line from the sun in a direction determined by the amount of the earth's angular motion in six weeks at that season, and measure off on it $64\frac{4}{5}$ equal parts. When a quarter of a year has elapsed the diameter will be $31\frac{8}{10}'$, and the line must contain $63\frac{6}{10}$ equal parts, its direction being known from the angular motion of the earth. By increasing the number of these observations, and joining the extremities of the proportional lines thus laid down, we shall have an ellipse with the sun in one of the foci. In the figure the eccentricity is much greater than is actually the case in the earth's orbit.

In learning the point of the earth's orbit in which the sun's apparent diameter is largest we have discovered its perihelion or point nearest the sun. In learning the point in which the sun appeared smallest we have determined the earth's aphelion or point farthest from the sun. These being known we may ascertain the amount of the eccentricity of the earth's orbit. The sun's greatest apparent diameter is $32' \, 35''.6$; its least is $31' \, 31''$; hence the radius vector at the aphelion : radius vector at the perihelion : : 32.5933 : 31.5167 : : 1.032 : 1. Half of the difference of the two equals the distance of the focus of the ellipse from the centre, a quantity which is the measure of the eccentricity of a planetary orbit.

§ **211.** As we have learned the variations of the earth's distance from the sun by noting the sun's varying size, so we may learn the varying speed of the earth's motion in her orbit by observing the variations in the sun's apparent motion. From what was said before it is evident that whatever motion in her orbit the earth has we attribute to the sun. If therefore the sun at the perihelion moves in twenty-four hours over an arc of 61′, while at the aphelion he describes in the same time an arc of only 57′, we must suppose these changes to take place in the earth's motion. Our varying distance from the sun accounts for a portion of these changes, for the angular velocity varies inversely as the distance. If it accounts for all then the raţe between the largest and shortest arc described in twenty-four hours would be the same as that of the largest and smallest apparent diameters. That is $\frac{61}{57}=1.07$, would equal $\frac{32.5933}{31.5167}=1.034$. But the first fraction is the square of the second, for $1.07=(1.034)^2$. Therefore the earth's angular velocities are to each other inversely as the squares of the distance at the perihelion and the aphelion. And this is found to be true of every part of the orbit.

If the angular velocities described by the earth in single days are inversely as the squares of the distances, the distances are inversely as the square roots of the arcs. Thus the relative distance of the earth from the sun in every point of its revolution may be easily calculated. Thus its perihelion distance is to its aphelion as $\sqrt{57}$ to $\sqrt{61}$, or as 1 : 1.034. The difference between the two is nearly $\frac{1}{29}$ of the perihelion distance, a quantity, as we shall see hereafter, no less than 3,000,000 miles. This variation of the earth's speed in her orbit thus discovered from observation is what we shall find must take place in elliptic orbits, and thus confirms the form of orbit which we have assigned to it.

§ **212.** In the earth's daily rotation we find no variation. Equal arcs of the equinoctial present themselves in equal times, proving that the earth's rotation is always equally rapid. If there were any jerk or sudden stoppage in the motions of the earth it would be perceptible to us by its effects. Even a diminution of her rotary speed would precipitate the Atlantic on the shore of Europe and lay bare the eastern shores of America. But no such violence is seen in the heavens. The impulse was from the beginning, and given by an all-powerful hand. All on the surface of the earth partakes its motion, and we are no more sensible of it than the fly who goes back and forth on the deck of a vessel is conscious of his motion down the stream. All around us seems at rest or moving only by its own motion. The bird who hovers over a field shares the motion of the air, or he would appear to us to move toward the west. Thus every motion we make is in reality combined with the rotary and revolving motion of the earth, the translation of the sun, and perhaps with the progression of our whole cluster, and even with other grand movements.

§ **213.** We will now inquire into the shape of the earth. Many circumstances obvious to the most careless observer give us an idea of its form. We will begin with these, and then from more exact experiments and reasoning deduce its precise form.

In the other planets and the moon we see only the globular form, we may therefore expect from analogy to find the earth a globe. The earth's shadow cast on the moon in an eclipse is always convex. Eclipses take place when the earth is in every variety of position between the sun and moon, and a body which whatever way it is turned casts a circular shadow must be a globe or sphere. A cone or a cylinder may cast a circular shadow when the sun is in the direction of its axis, but a shadow which is circular on whatever side it is projected must come from a sphere. When on the ocean a ship comes in sight, at first only the top of her mast and gradually the whole mast and hull rise to view; and as the ship sails away the hull, mast and yards successively disappear. As we approach land the mountains, steeples, houses, and lastly the shore itself becomes visible; as we sail away, they gradually sink away at the base. This may be seen in all parts of the ocean, because owing to the mobility of water the parts of the globe covered with water keep their normal shape. It may be seen in extensive plains or deserts; but in most parts the surface of the earth is so rugged or so interrupted by minor objects that we are less sensible of these phenomena. This irregularity, however, is too slight to affect the general surface of the globe; the highest mountains are but five miles above the level of the sea, and their mass no more takes off from the earth's rotundity than particles of dust from the form of a celestial globe. These phenomena occur every where and terrestrial objects rise and vanish with equal rapidity in whatever direction we approach or leave them; thus proving that the earth has a convex surface which falls away equally or nearly equally in all directions.

§ **214.** Hence in constructing canals an allowance of eight inches a mile below the horizontal plane is necessary, to compensate for the curvature of water at rest. And if we imagine a large portion of the ocean which now appears a plane surface to be frozen, and to be cut off, a slice which is two miles across would rise eight inches in the middle. Rivers flow to the sea because their origin is above the normal curve of the earth. They indicate by the slowness or rapidity of their course the slope of the country they traverse, that is the excess of their curve over the curve of the sea.

Another proof of the earth's globular form is the testimony of those who have sailed round her. She is of moderate size and has been circumnavigated in more than one direction, and navigators assure us that they have every where observed the gradual appearance and disappearance of objects, and every where their view has been bounded by a circumference about three miles distant. This extent included within the visible horizon is a slice of the earth, and the reason no greater extent is visible is that a globe the size of the earth falls away so much that it is out of sight beyond this bounding line. On a smaller globe the surface would fall away more and the terrestrial horizon would be less extensive, on a larger globe we should have a wider horizon.

Now the only body from which none but circular sections can be made is a sphere; the earth therefore must be a sphere or nearly a sphere.

§ **215.** Before inquiring into the exact form of the earth we must make ourselves acquainted with another set of circles and poles which like the ecliptic and equatorial with their secondaries, serves to define the place of a star. I speak of the celestial horizon, a circle every where 90° distant from the zenith or pole over our heads and from the nadir or pole beneath our feet. This great circle is not immovable and the same to all the inhabitants of the globe like the equinoctial and ecliptic, it differs with the position of each observer. We make our own horizon. By advancing north or south, east or west, on the globe, we change the point over our heads and consequently the boundary line 90° distant from that point. If we stand still, the earth in her rotation presents us each moment with a new horizon, and each night, owing to her revolution, she presents us a horizon partially different from the one we saw at the same time the night before. It is often more convenient to refer heavenly bodies to the horizon and its secondaries then to the equinoctial. The observed distances can afterward be converted into the declination and right ascension or the celestial latitude and longitude. The distance of a star above the horizon is called its altitude; it is measured on a vertical circle. The body may also be referred to the zenith by a vertical circle, and the arc intercepted is called its zenith distance. The zenith distance is the complement of the altitude.

§ **216.** Azimuth is the angular distance of a celestial object from the north or south point of the horizon (according as it is the north or south pole which is elevated), when the object is referred to the horizon by a vertical circle. Or it is the angle comprised between two vertical planes, one passing through the elevated pole, the other through the object. When the body is on the horizon it is only necessary to count the number of degrees between that point and the meridian in order to find its azimuth. But if the point is above the horizon, its azimuth is estimated by passing a vertical circle through it and reckoning the azimuth from the point where this circle cuts the horizon. The altitude and azimuth of an object being known, its place in the visible heavens is determined.

We may learn the zenith and nadir by direct observation, or we may learn the horizon and find the zenith and nadir, each 90° distant from it. The nadir is always in the direction of a plumb line, the zenith 180° from the nadir. The zenith and nadir are the vanishing points of all lines in all parts of the earth mathematically parallel to the direction of a plumb line at the observer's station.

The celestial horizon is the vanishing line of a system of planes parallel to one another, passing, one through the centre of the earth, another through the place of the spectator, another through the point on the earth's surface opposite to him. All these planes cut the concave sphere in one and the same line, for the earth included between them is no more to them than a grain of sand between two circles of paper. The edges of the paper coincide and so do all the parallel planes which can be drawn between one extremity of the earth and the other, if they are only continued sufficiently far.

§ **217.** This great circle thus marked out in the heavens, and passing through the centre of the earth is called the celestial or rational horizon. It divides the sphere into two equal parts. But as we do not stand at the centre of the earth we use the plane parallel to this and touching the earth where we stand, called the sensible horizon. The sensible horizon is parallel to the rational, and distant from it by the earth's radius, or 4,000 miles. As we have above said, it coincides with the rational horizon, of course it divides the sphere into two equal parts. The rising and setting of stars are the same to the sensible and the rational horizon. The sun rises a very little and the moon a little earlier, and both set later to the rational than to the sensible horizon. The horizon or bounding line which to our eye unites earth and heaven is a different circle from these. It is not a great circle of the earth, and consequently when prolonged not a great circle of the heavens.

An eye perfectly level with the surface of the sea would

have no visible terrestrial horizon, and would see that portion of the heavens above the sensible horizon. Let the eye be raised even to the height of a man, five feet, and a segment of the earth's surface, nearly three miles in every direction becomes visible. The bounding line of the segment is our visible horizon. It is not in the same plane with the sensible horizon, but below it. When prolonged it cuts the concave sphere below the sensible and rational horizon. This depression of the visible horizon below the direction of a spirit level is called the dip of the horizon. It must always be allowed for, as observers are always more or less raised above the level of the sea. To get the true altitude of a star above the sensible horizon, the angle of depression must be subtracted from the observed altitude.

§ **218.** If the earth were a larger sphere, a portion of its surface would fall away less from a plane, and the dip of the horizon would be less. On a smaller globe the depression would be greater, and more of the heavens below the sensible horizon would be brought into view. Knowing the amount of the dip, the falling away of the earth in a given portion of her circumference, and consequently the convexity of her surface, her size may be calculated; for a given convexity can belong only to a sphere of a certain size.

As we ascend elevations, though a much greater expanse of country is exposed to our view, not only is each object smaller but the whole picture appears smaller, because it is included within a smaller angle. This may be seen in an instant by opening a pair of dividers so as to embrace part of a globe. When they are nearly wide open, they will include a very small portion of the globe, but the eye of the spectator at the centre of the dividers would see this portion under a very large angle. If the dividers are closed more and gradually lifted, they will embrace more of the globe, but the visible portion will be seen under a smaller angle. In like manner a movable tangent passes from the eye of the spectator in every direction forming his horizon. As he ascends, this tangent reaches to a greater distance on the globe, but forms with itself a smaller angle.

§ **219.** Let us now consider what effect elevation has on this visible plane considered as a circle of the heavens. If the earth were cut in halves and a man stood in the centre of its plane surface his height would be so trifling compared to the surface on which he stands, that his visible and rational horizon would agree. The sun, moon and stars would rise over the rational and visible horizon at the same moment. But as man stands on a pinnacle formed by the convexity of the earth, it is the same as if part of the earth were hewn from under him and no longer kept out of sight so much of the heavens. The more it is hewn away, that is the more convex it becomes, the more of the starry sphere is visible, for here, as in the moral world, only the opaque earth hides the heavens. In considering this effect of the earth's convexity it is better to compare the visible with the sensible than with the rational horizon; the effect of the earth's convexity in increasing the visible portion of the heavens will be more apparent than if we use the rational horizon. We see its effect daily in the lighting up of mountain tops long before and after valleys are shadowed. Aeronauts tell us they have seen the sun set three or more times in one evening. This was occasioned by repeatedly increasing their elevation, and thus again bringing the sun above their horizon. On the Peak of Teneriffe, a mountain 13,000 feet high, Humboldt found the surface of the sea depressed on all sides nearly two degrees. The sun rose to him twelve minutes sooner than to an inhabitant of the plain; and from the plain the top of the mountain was seen enlightened twelve minutes before the rising and after the setting of the sun.

§ **220.** The dip of the horizon at different elevations may be observed; or the shape and dimensions of the earth being known, it may be calculated, for different elevations. When the spectator's eye is one foot above the sea-level, the dip is 59″, when 100 feet, it is 9′ 51″. As almost all land is raised above the level of the sea, and as observations are usually taken in high places, tables have been constructed showing the dip of the horizon at all required elevations.

The reason that the observer sees only the segment of

the earth's surface which is bounded by the visible horizon is that the earth falls away so suddenly that the edge of what is seen conceals all beyond. If the earth were a larger sphere a given amount of elevation would make a larger ring-shaped surface visible than now becomes so; if it were a smaller sphere, the same amount of elevation would bring up a narrower ring below the natural horizon. As it is the visible horizon enlarges rapidly as the observer mounts. An eye placed five feet above the surface of the sea sees 2¾ miles every way. If it be elevated twenty feet, that is to four times the height, it will see 5½ miles, or twice the distance. At a height of 100 feet the horizon is 13 miles off.

A much larger portion of the atmosphere than of the earth is visible to us. The greatest distance of the clouds in the horizon at sea is 94 miles in every direction from an observer. Consequently the whole extent or diameter of the horizon is 188 miles; and the circumference is 590.97 miles. Thus the physical visible horizon extends only 2¾ miles; if we look higher it extends 94 miles; if we take in the fixed stars it is millions of millions of miles.

Probably the greatest extent of surface even seen at once by man was in the aeronautic expedition of Biot and Guy Lussac. They were elevated nearly five miles above the surface of the earth. Now the convex surface of a spherical segment is to the whole surface of the sphere to which it belongs as the versed sine or thickness of the segment is to the diameter of the sphere. Its thickness in this case almost exactly equals the perpendicular elevation of the point of sight above the surface. The proportion of the visible area to the whole earth's surface would in this case be that of five miles to 8,000 or of one to 1,600. The portion visible from Mount Etna or the Peak of Teneriffe is about one four thousandth.

Having found rudely the shape of the earth, we will, before we attempt to ascertain its precise shape, learn its actual size. This is of the utmost importance, for it is one unit in calculating the size and distance of the heavenly bodies. As we cannot embrace in one view any large portion of the earth, nor can we retire to a distance and view it as a whole, all the knowledge we can have of it is gained by exact measurements of small portions of its surface accompanied by geometrical deductions. If we were sure the earth was a sphere we could at once tell its precise dimensions. If we can learn that it is a spheroid, the process will be somewhat more difficult.

§ **221.** We may ascertain the earth's diameter by measuring the heights and distances of two stationary points which can barely be discerned from each other, and but for the effect of refraction, by which we are enabled to see a little round the interposed segment of the earth's surface, this method would be pretty correct. It is known from observation that two points each ten feet above the earth's surface are visible from each other over still water at a distance of nearly eight miles. This, by a simple calculation, gives 8,450 miles for the diameter of the earth.

Another method, which is the one usually adopted, is to measure the length of a degree of the meridian. Thus let the latitude of the place g be determined with great accuracy by repeated observations of the heavenly bodies; determine also the latitude of h; for simplifying the process, we will suppose the two places to be exactly north and south of each other. If the distance between them be measured with great care, a simple proportion will give the approximate circumference of the globe; thus, supposing the latitude of h to be 50° 54′ N., and that of g to be 56° 24′ N., and the measured distance between g and h to be 390 miles. Then 5° 30′ : 360° : : 390 miles : 24,900 miles=the circumference of the globe.

§ **222.** We may learn the shape of the earth by studying phenomena at its surface, or by considering whether the attraction of the earth on the moon is that of a sphere or a spheroid.

The simplest though not the most exact way which presents itself is to ascertain the direction of the surface as it would exist without the accidental irregularities of hills and mountains; and this is determined by means of a plumb line, the line always taking a direction perpendicular to the surface. The plumb line is to be sure slightly diverted from the centre of the earth by the attraction of mountains, and sometimes by dense subterraneous matter to a greater degree; but this difficulty may be obviated by multiplying experiments.

If the earth is a perfect sphere plumb lines let fall on all portions of its surface will converge to one point. If not they will converge along certain curves in the interior of the earth. Three methods have been employed to investigate the curvature of the earth's surface. It has been inferred from the measurement of degrees in different places, from the vibrations of a pendulum, and from certain inequalities in the moon's orbit. The first method is directly geometrical and astronomical. In the two others we infer from movements accurately observed, the measure of the forces which cause those movements; and from the inequality of the forces we infer the difference between the equatorial and polar diameters. Conclusions respecting the figure of the earth, founded on the increase of the attracting force in going from the equator to the poles, are dependent on the distribution of the density of the interior. Its ellipticity may be calculated on the supposition that the earth does or does not increase in density toward the centre. A comparison of the earth's figure with its velocity of rotation makes an increase of density toward the centre probable, and a comparison of the ratios of the polar and equatorial axes of Jupiter and Saturn with their times of rotation shows the same increase to exist also in these planets. Actual measurements calculated on the supposition that there is this central density, give very nearly the same compression as the measurement of degrees and the moon's inequalities.

§ **223.** If the earth is not a perfect sphere some portions of its surface must be nearer to the centre than others, consequently gravity must act more strongly in these portions than elsewhere. The most delicate measure of the force of gravity which we have is the swinging of a pendulum, because this shows on a large scale the effects of gravity. A pendulum is a weight suspended either by a line or by a bar of wood or metal. If the weight is drawn aside from the perpendicular, gravity draws it back to that line, and the momentum thus acquired causes it to describe an arc on the other side of the perpendicular. If the arc of vibration is small the times of vibration are equal as long as the motion continues.

There is a constant relation between the lengths of pendulums and the times of their vibrations ; the same relation which exists between the spaces and the times of bodies falling with accelerated motion. The squares of the times of vibration are in proportion to the lengths of the pendulums.

If two pendulums be made of the same length and one be carried to the equator, the other near the pole, and it is found that one of these vibrates more rapidly than the other, we must suppose it influenced by a more powerful attraction. At the equator the vibration of a pendulum is slackened, showing that gravitation is weaker there ; near the poles vibrations are more rapid, showing increased gravitation. We infer therefore that we are nearer the centre at the poles than at the equator, and that the earth is not a sphere but a spheroid. By comparing pendulums of the same length, and marking the number of vibrations made by them in equal times in different latitudes, the amount of gravitation in these places is ascertained. The intensities of the forces will be as the squares of the numbers of vibrations at the two places. A pendulum which under the equator makes 86,400 vibrations in a mean solar day, transported to London, makes 86,535 vibrations in the same time. Hence gravitation at the equator as to that at London as $86{,}400^2 : 86{,}535^2$, or as $1 : 1{,}00315$.

§ **224.** Experiments made in all accessible latitudes give $\frac{1}{194}$ for the difference in gravitation at the equator and at the poles ; the weight increasing (from causes we shall presently explain) as the square of the sine of the latitude.

Since the weight of all bodies is affected in the same proportion it is not easy to prove this increase of gravitation directly. The weights we would use are as much increased toward the poles as that which we would balance with them. What we call a pound at the equator, is at the pole more than a pound, though we still call it a pound. The pressure exerted on the hand by a mass weighing 194 pounds at the equator, would be at the poles equal to the pressure of a mass weighing 195 pounds at the equator. This has been thus explained. Imagine a weight x suspended at the equator by a string without weight passing over a pulley and conducted, if such a thing were possible, over other pulleys till the other end hangs down at the pole and there sustains the weight y. If then the weights x and y are such as at either station, equatorial or polar, would exactly balance one another, they would not in this supposed situation balance one another, but the polar weight would preponderate; to restore the balance the weight x must be increased by the $\frac{1}{194}$ of its mass.

§ **225.** Let us consider whether all the increased gravitation shown by the pendulum as we approach the pole is owing to the increased convexity of the earth, or whether some other cause may exist. The rotation of the earth immediately suggests one, in the centrifugal force created by it. This centrifugal force is greatest at the equator, because the equatorial particles revolve with more rapidity than any others. Moreover the direction of the centrifugal force communicated by rotation is always opposite to the direction of the radius of the different circles in which the particles move. Now gravity or the centripetal force is always in a direction perpendicular to the surface, and therefore it is only at the equator that this is exactly in an opposite direction to the centrifugal force. The quantity of force directed from the surface (for strictly speaking it should not be called centrifugal force) at any point not in the equator may be resolved into two forces, only one of which is directly opposed to gravity, and is the true centrifugal force. The other is perpendicular to the direction of gravity, and does not diminish its force. Thus only in the equatorial regions does the whole amount of centrifugal force take effect in diminishing the force of gravity. In parts distant from the equator a portion only of the centrifugal force, and that in its diminished state, acts in opposition to the force of gravity. The diminution of gravity at the surface of the earth arising from the centrifugal force varies as the square of the cosine of the latitude.

§ **226.** This centrifugal force at the equator lightens bodies by $\frac{1}{289}$ part of what they would otherwise weigh there, and by $\frac{1}{288}$ of what they would weigh at the pole were the earth a perfect sphere. At the poles the gravitation of a body equals the whole force of gravity at that distance from the centre, at the equator it equals the force of gravity at that distance minus the centrifugal force. We must therefore, in accounting for the variations of the pendulum, attribute one part of the change to the unequal convexity of the earth, and one part to the inequality of centrifugal force in different latitudes. Of these two causes of diminution the centrifugal force is the more powerful. It diminishes gravitation by $\frac{1}{289}$, while the elliptic form of the earth diminishes it only $\frac{1}{590}$. The two together make up the difference shown by the pendulum $\frac{1}{194}$. The earth's size and velocity of rotation being known, her ellipticity and the consequent increase of gravitation at the poles, and also the diminution of gravitation at the equator, arising from the centrifugal force, are matter of pure calculation. Since the effects calculated for those two causes equal the observed difference the assigned causes are the true ones.

The rate of increase of gravitation shown by the pendulum, in travelling from the equator to the pole, is as the square of the sine of the latitude. Gravitation would be inversely according to the squares of the radii if the earth were at rest. It is diminished in each latitude by centrifugal force. Centrifugal force in each latitude, as we have shown, is according to the squares of the cosines; therefore the actual gravitation in each latitude is as the square of the radius minus the square of the cosine, that is, as the square of the sine.

The earth's polar diameter is to its equatorial as 298 : 299; or more correctly the polar diameter is 7,899.171, and the equatorial 7,925.648 miles. For the convenience of round numbers it is usually called 8,000 miles. The difference between the polar and equatorial radii is thirteen miles. The excess of the equatorial radius is $4\frac{3}{7}$ the height of Mount Blanc, or $2\frac{1}{2}$ the probable height of the Dhawalagiri, in the Himalaya chain.

§ **227.** We will now see what information concerning the earth's shape and size we can obtain from the heavens. Since the earth affords us no marks by which to trace or estimate our course we must look to the stars to ascertain our progress and to guide our path. The polar distances of the stars are known; therefore by observing their meridian altitudes we can learn the height of the pole and consequently our own latitude. When our latitude has diminished a degree we know that, provided we have kept to the meridian and the earth is a symmetrical figure, we have described one three hundred and sixtieth part of the earth's circumference. By observations on the polar star we may keep the true direction of the meridian even if local difficulties oblige us to turn aside from it for a while. Owing to the irregularity of the earth's surface, which often makes it impossible to travel due north or south or east or west, and which constantly changes the plane of the observer, it is exceedingly difficult to determine the length of a degree of a great circle of the earth. Of course it is not necessary to measure a whole meridian, a few degrees in known latitudes are sufficient, and calculation will then give the true form of the whole. Neither is it necessary to measure precisely a degree, we need only know exactly how much, be it more or less, we have measured. We determine by astronomical observations the exact difference of the latitude of the two stations, and measure this distance on the ground accurately. It is of great importance to avoid even slight errors in these two operations, for an error committed in a single degree will be magnified 360 times in the circumference and nearly 115 times in the diameter of the earth concluded from it. The altitude of the observed star also, since it affects the latitudes, must be found very accurately. The true place of stars which are near the zenith can be found more precisely than the place of those which are nearer the horizon. Instead therefore of observing stars which may be nearer the horizon, a star near the zenith is observed, and if its zenith distance is raised or depressed a degree we know that the pole must also have been raised or depressed a degree, for the polar distance of the star cannot have altered. If then at one station a star passes through the zenith and at another it passes one degree north or south of the zenith, we are sure that the latitude of the places differ by the same amount.

§ **228.** If all measurements of a degree on all great circles of the globe agreed we should have no hesitation in calling the earth a sphere. If in travelling in one direction round the sphere we should find that equal portions of the celestial sphere rise to equal measured distances of advance we should infer that the earth is symmetrical in that direction. If travelling in a second direction for equal advance on earth unequal portions in the heavens rise, we shall infer that in this direction the earth differs from a sphere.

There is but one great circle of the earth, equal portions of which correspond to equal portions of the celestial sphere; this is the equator. When a man on the equator advances sixty geographical miles, his horizon (allowing always for the earth's rotation) will advance one degree on the celestial sphere. For each of the 360° of longitude he will see successively an additional degree of right ascension. If a man traverses the earth from north to south in the line of a meridian, he no longer finds that for equal distances travelled his horizon advances equally. At the equator 68.73 statute miles north or south bring up a degree of the heavens. In latitude 43°, 68.99 are required; in 80° lat., 69.36; and at the poles 69.39 miles are required to bring up one degree. The more rapidly a surface alters its curve, the more does its tangent vary at each observation, the less distance need we go to vary it a certain amount, for instance one degree. Comparing observations in longitude and in different latitudes we infer that the equator is the only great circle of the earth, and is larger than any other curve on its surface; and likewise that the curvature of the earth is greatest at the equator, and consequently that the meridians, though called circles, are in reality elliptic curves; and are shorter than the equator.

We must apply the same correction to the form of the visible actual horizon formerly considered as round. It can be perfectly circular only at the poles. At the equator it must be elliptical, and longer from east to west than from north to south. Between the poles and the equator it assumes various forms composed of oval curves having the shorter axis running north and south.

To determine the exact amount of the earth's ellipticity requires nice observations in different latitudes.

§ **229.** More than a dozen measurements of degrees, most of which belong to this century, have now made us quite accurately acquainted with the dimensions and the ellipticity of our globe. Only two of these measurements have been throughout their whole extent actual and mechanical. The irregularity of the surface of most countries renders this mode inadvisable. The first of these, made in 1635, was of that arc of the meridian which lies between London and York. The difference of the latitudes of these cities was first ascertained. This gave the number of degrees in the arc to be measured. The distance between the two cities was then actually measured, and the turnings and windings of the road, and the ascents and descents were afterward allowed for. This measurement gave the length of a degree too large by 1,000 yards.

The only other instance of the actual measurement of an arc is that made by Mason and Dixon. They measured with rods a line of nearly 100 miles, in Pennsylvania, near latitude 39° 12′.

§ **230.** A more accurate mode of finding the length of a degree is by a combination of actual measurement, and of trigonometrical operations founded upon it. Two places are selected which lie under the same meridian or nearly so, and the difference of their latitudes, which gives the number of degrees of the arc to be measured, is ascertained with the utmost precision. A base line of a few miles of extent, and at some little distance from the meridian arc is then very carefully measured; this is the only actual measurement which need be made. The extremities of this base line are then connected with the extremities of the meridian arc by imaginary triangles, the sides of which are not measured, but trigonometrically determined from the length of the first base line and the angles of the triangles. A theodolite measures the angles accurately. The stations from which the angles are observed should be so selected that none of the angles shall be very small, otherwise a slight mistake in an angle causes a great error in the opposite side of the triangle. Toward the conclusion of the process one of the sides of one or more of the triangles is measured and its length compared with that found by computation. This base of verification is taken as far distant from the first base as circumstances will admit. In one of the French operations the base of verification was between four and five hundred miles distant from the first base, and was seven miles in length, and yet the difference between its computed length and that obtained from its actual measurement did not amount to twelve inches.

§ **231.** The following particulars show the accuracy which distinguishes these operations and the means taken to ensure it. A base of five miles in length was measured in Hounslow Heath with a steel chain of exquisite workmanship. The same base had been measured three years before with glass rods, and the two measurements differed only 2½ inches. Sometimes rods of platina or of iron are used for measuring, and an allowance is made for the changes of temperature affecting the rods in the course of the operation. In later measurements rods composed of different metals put together so as to show the slightest contraction or expansion have been used.

In 1735 two scientific expeditions were sent from France, to determine the length of a degree of longitude in different latitudes. One degree was to be observed upon the equator, the other as near the poles as possible. One degree was measured in the valley of the river Tornea in Lapland. The base was measured on the frozen surface of the river, with a view to obtain as level a plain as possible; and rods of deal were employed instead of metal on account of the extreme cold. Two independent measures by two sets of observers differed only four inches. These operations were completed several years before the return of the Peruvian expedition, which had to contend with extraordinary difficulties caused by the ill-will and indolence of the natives, and by the localities. Their station was a mile and a half above the level of the sea, and in some instances the heights of two neighboring signals differed more than a mile. To accomplish their measurements occupied nine years, three of which were employed in the determination of latitudes alone.

At the beginning of the French revolution, a measurement was made from Dunkirk to Barcelona, in order to ascertain the length of a quadrant of a meridian, and take the ten millionth part of it as a metre or universal standard. A metre contains 39.37 inches. This method of obtaining a standard of measure is not so good as the English mode, which consists in observing the length of the pendulum, which in a certain latitude, (that of London,) in a vacuum at the level of the sea beats seconds of mean time. The length of the pendulum is ascertainable without the use of any linear measure whatever ; whereas in determining the French standard or the quadrant of the meridian, some linear measure already in use must be employed. And thus the very basis of their new system is expressed in terms of that in the place of which it is substituted. Arcs of longitude have also been measured in various other latitudes, and their observed lengths agree with their theoretical lengths.

§ **232.** By measuring degrees of the meridian we obtain the compression of the earth, while pendulum observations give us the ellipticity confounded with the effects of centrifugal force. The motions of the moon confirm the shape assigned to the earth. They cannot be accounted for on the supposition that the earth is a sphere, but they agree perfectly with the supposition that it is a spheroid. The ellipticity inferred from the lunar inequalities has an advantage not possessed either by measurements of degrees or by pendulum experiments, in being independent of local accidents, and thus showing the mean ellipticity of the earth.

CHAPTER XI.

GENERAL PHENOMENA ON THE EARTH'S SURFACE.

Universal diffusion of Gravity over the Earth's Surface. Determination of the Earth's Mass and Density. Our Knowledge of the Earth's Surface. The Sea. Tides. Stability of the Ocean's Equilibrium. The Atmosphere. Clouds. Winds. Trade Winds. Use of the Atmosphere. Absorption and Diffusion of Light and Heat. Refraction. Twilight.

§ **233.** The earth's exact diameter is important as a unit of measure for other bodies; its density is no less so as a standard of comparison for other planets.

The powerful attraction which the earth, in consequence of its superior bulk, has for bodies on its surface, prevents our perceiving the attraction they exercise on one another. It may be shown however by balancing a small mass in such a manner that it may be moved by the slightest influence, and then bringing a large body into its neighborhood; or by ascertaining the deflection of the plumb line caused by the vicinity of a mountain; or by comparing the length of a pendulum vibrating seconds in a plain, and on the summit of a mountain. A balance of torsion is in fact a horizontal pendulum. It may be applied to a mountain or to much smaller attracting bodies. If two equal balls of lead are suspended from the opposite extremities of a slender bar of wood, and this is suspended at its centre by a very fine wire, the only force required to move the balls will be that which suffices to produce a slight twisting of the wire that suspends the rod. Now if a large mass of lead be brought into the neighborhood of each ball, (the rod having been previously hanging at rest,) its attraction will cause the rod to turn round, until the small balls have come into the same line with the large masses. If the masses be now moved a little further, the balls will follow them; twisting the wire from which the rod is suspended still more. Now, as the force which is required to produce any amount of alteration in the position of the rod can be ascertained in another way, the actual amount of the attraction which the masses exercise over the balls may be determined; and this may be compared with the earth's attraction. From the knowledge of these facts, the quantity of matter in the earth may be compared with that in the masses of lead; for the weight of the earth is just as much greater than that of the masses of lead, as the force with which it attracts the balls exceeds that with which the masses attract them, proper allowance being made for their difference of distance. When the actual weight of the earth is known, we may estimate its density as compared with water; since we may easily calculate the weight of a globe of water of equal size. And from the weight and density of the earth, that of the other planets and of the sun may be ascertained.

§ **234.** A beautiful proof that the attraction of gravity is diffused through separate portions of the earth is the fact that mountains draw a plumb line out of the perpendicular.

By ascertaining the exact amount of the deviation and obtaining the specific weight of the mountain, the specific weight of the earth, or its weight compared with the mountain and consequently with a globe of water of the same size, may be learned. The mountain Schehallien in Scotland was thus examined. It was measured from its base to its summit, its component parts examined, and its specific gravity determined. Assuming that the spirit in the levels of the instruments would be attracted toward the mountain, or that the plumb lines by which the instruments are rectified would deviate from a perpendicular to the horizon, it is plain that observations on the fixed stars, taken on opposite sides of the mountain, would differ from each other by double the amount of deviation.

§ **235.** The meridian zenith distances of certain stars were observed first on the north and then on the south side of the mountain. They gave a constant error of $11\frac{1}{2}''$ more than could be accounted for by the difference in latitude of the stations. The mountain therefore deflected the plumb lines from the perpendicular $5\frac{3}{4}''$. From the actual attraction of the mountain its attraction at the distance of the earth's centre was to be calculated. The comparative powers of attraction of the earth and the mountain, and their relative sizes being known, their relative densities could be determined. After a year's labor in reducing these data it was found that the density of the earth was to that of the mountain as five to three, or nearly five times the density of water, or nearly double that of rocks near the surface. It is of about the density of silver ore throughout. Since the density of the earth is so much greater than the average specific gravity of rocks on the surface, it follows that the internal matter must be more dense than the superficial layers.

§ **236.** While we thus by reasoning learn the size and mass of the earth, how little do we actually know of it. Even its surface is not yet wholly known to us. The ancients thought a fiery impassable zone separated the northern from the southern regions; the polar circles have proved equally impenetrable to the most zealous efforts of the moderns. The interiors of the continents are yet unexplored, so that enough remains to stimulate the curiosity and enterprise of man for ages yet to come. Enough of the surface has been explored to prove to us that its nature is every where the same. Every where there are traces of convulsive change; every where fire and water leave their traces. Huge rocks have been melted and cast up; water has worn them away and left the sand and the pebbles to tell of the slow destruction it has wrought. To water also we owe all the plains and habitable spots of the earth. It is still busy bringing all things to its own level. All over the globe there is no sameness, each country has its characteristic scenery, every where there is variety. Even the bottom of the sea has its risings and its abysses, fit abodes now for its varied inhabitants, and perhaps the peaks and valleys of a world to be upheaved hereafter. But beyond this very outer surface we cannot penetrate. The deepest mines are but as a scratch on the surface of a model globe, the highest mountains are not five miles high. The ocean has not been sounded below 27,600 feet. The deepest mines do not penetrate more than 2,231 English feet, or $\frac{1}{9800}$ of the earth's radius below the level of the sea. The absolute depth of mines often exceeds this, for they are usually situated in elevated valleys or mountains; sometimes the absolute is less than the relative depth. By studying the edges of strata which dip and rise again at a distance, and by observing their dip we can learn with absolute certainty the depth of the basin formed by them. We can thus infer the nature of the coast six thousand feet below the level of the sea. By adding these depths to the mountain summits, we have 48,000 English feet or $\frac{1}{524}$ of the earth's radius known to us. Loosely infolding the hollows of this surface lie the waters of the ocean. Gravity keeps them in their place, gravity swells them and gives them the true form of the earth, the normal form which it would retain all over were it not for the rigidity of the rocks.

§ **237**. If we would learn the general form of the earth's surface we must seek it in the most mobile substance, for this obeys no minor principles of stratification, but the great forming principle of gravity. The ocean gives us the form of the globe when not interfered with by the hardness of materials. It also shows us that this form is liable to change. Twice every day some parts of the globe's outline increase, others diminish in convexity.

The moon is so near to the earth that the difference between the earth's surface and its centre bears a considerable ratio to the moon's distance; it is 4,000 out of 240,000 miles. The waters next under the moon are therefore as much attracted as the centre of the earth and more also. It is this excess of the moon's attraction which causes the tide. As the sun is so much more distant the earth's radius bears to his distance a smaller proportion; the difference between his attraction on the earth's surface and at its centre is less. Thus although the sun is so much larger than the moon it affects the tides only $\frac{2}{5}$ as much.

In obedience to gravity the waters nearest the moon fall towards the moon and form tides. The waters nearest the moon are of course less distant from the moon than the solid earth is; these particles move easily, they are drawn up. The earth also is attracted, but not so much. The waters most distant from the moon are still less attracted and they remain more distant. Thus the waters in a straight line with the moon bulge from the earth's surface; part of them because they are drawn from the earth, part because the earth is drawn from them. Opposite sides of the earth always have similar and equal tides at once, whether high or low. The sun and moon act in the same direction when the moon is new or when it is full. In both cases the tides are at their highest, and are called spring tides, and occur twice a month. When the moon is in her quarters the sun and moon act at right angles to one another, the tides are low and are called neap tides. The solar wave is lowest when the lunar wave is highest, and the solar highest when the lunar lowest.

§ **238.** The point of the earth's surface to which the moon is vertical is the highest point of the waters; 90° in every direction from this the waters are drawn away in consequence of the moon's attraction and the limited quantity of water; and to all such places it is low tide. Since every meridian is under the moon and opposite to it in the course of twenty-four hours, it is high tide twice and low tide twice in twenty-four hours. In those latitudes to which the moon is never vertical the tides cannot be so high as in those nearer the zodiac. Within the polar circles their ebb and flow are scarcely discernible. The highest and lowest tides occur in March and September, because then the sun and moon are in or near the same plane.

Since the moon advances 13° daily in her orbit, the earth must not only make a diurnal revolution but advance 13°, before the midnight moon will be seen on the same meridian as on the preceding night. To accomplish this takes about fifty minutes. The mean duration of the ebb and flow will then be half of 24h. 50′, or 12h. 25′; and in one flow of the twenty-four hours the place will be under the moon, the next high tide removed from it by a space equal to more than the diameter or the semi-circumference of the earth.

§ **239.** The maximum height of the tide is not however when the moon is in the meridian, but about three hours afterward. In consequence of the impulse given to the waters, which indeed is only gradually diminished, they continue to flow, and thus accumulate for some time. In like manner the highest spring tides do not occur at the instant of the sun and moon's acting on the waters in conjunction, but sometime afterward.

The rising tide is called flood, and the falling, ebb tide. The average height of tide for the whole globe is about $2\frac{1}{2}$ feet. If the earth were covered uniformly with a stratum of water the difference between the two diameters of the oval would be five feet. Local causes sometimes give the tide a height of 70 or even 120 feet. The Atlantic on the shores of France and also of North America pours into the rocky shore with prodigious violence, while the gentle tides of the Pacific are scarcely perceptible. The winds have great influence on the height of the tides according as they conspire with or oppose them. But the actual effect of the wind in exciting the waves of the ocean probably extends very little below the surface. Even in the most violent storms the water is probably calm at a depth of ninety to a hundred feet. Currents and headlands modify the force and the time of the tide for each place, and in land-locked seas like the Mediterranean and Baltic there are no tides. If the globe were covered with islands and continents, leaving no large open sea, there would be no regular tide. The variations in the sun's distance are too slight to influence the tides much, but those in the moon's distance have considerable effect. When the moon is near at the time of the equinoxes the very highest tides are produced.

§ **240.** The equilibrium of the ocean can never be destroyed by the causes which influence the tides; they are sufficient to keep the ocean in perpetual agitation, but can never detach it or in any great degree withdraw it from the earth. The sun's influence on the ocean is only $\frac{1}{38448000}$ of gravity at the earth's surface, and the action of the moon is little more than twice as much. Had the action of the sun and moon been equal there would have been no neap tides. In ports where the tides arrive by two channels of lengths corresponding to half an interval, there is neither high nor low water, on account of the interference of the two tidal waves.

While our place on the earth is at the level of the ocean of waters it is at the bottom of the aerial ocean. This rests on the waters and fills up the valleys as the sea fills in its bed, and probably presents to observers in other planets the level, swelling, outline surface which the sea presents to us. In it man and animals move, by this they support life. All that meets their sight and hearing passes through this medium.

The atmosphere is an elastic fluid of great rarity, and vastly more compressible than water. Water being but slightly compressible is of very nearly the same density at all depths. Air being exceedingly compressible, the lower layers are pressed down by the weight of those above, and the density is much greater in the lower than in the upper regions. A column of air reaching down to the centre of the earth and compressed between walls, a few miles beneath the surface of the earth would be dense as gold. The lowest part of a similar water-column would have three million times the density of common water, and 119 times that of most marble. The pressure of the air on every square inch is fifteen pounds; so that the whole globe sustains a weight of 14,449,000,000 hundreds of millions of pounds.

The atmosphere contains a definite amount of air, as the ocean does of water. This is held to the surface of the earth and the sea by gravity. Its elasticity contends against gravity, and constantly endeavors to enlarge its volume and thus remove it farther from the surface of the earth. The lower layer is pressed on by all the rest; its elasticity is therefore less available, and it is more dense. The next layer above has less pressure from above, and is less dense. As the heights above the sea increase in arithmetical progression, the density of the atmospheric column diminishes in geometrical progression, consequently at a much more rapid rate.

§ **241.** By calculation founded on this proportion we learn that the atmosphere extends to about forty-five miles above the surface of the earth, a distance no greater compared with the size of the earth than the skin of a peach compared with the fruit within. This may not be the exact height of the atmosphere, but it extends at least as far as this, because at this height it reflects light, and it is evident that there must be some limit where elasticity and gravity balance one another. The height of that portion of the atmosphere which is sufficiently dense to reflect light may also be found rudely from the length of time twilight lasts.

We may learn from the barometer the proportion in which the atmosphere is distributed. This indicates, at 1,000 feet above the level of the sea, that we have left below us one thirteenth of the mass of the atmosphere. At 10,600 feet of perpendicular elevation, (which is rather less than that of the summit of Etna,) we have ascended through about one third. At 18,000 feet, (which is nearly that of Cotopaxi,) we have beneath us one half the atmosphere in weight. At an altitude not exceeding the hundredth part of the earth's radius the thinness of the air would be so extreme that neither combustion nor animal life could be maintained. The atmosphere, however, only extends to about the eighteenth part of the distance of a radius. Its mass is not more than $\frac{1}{100000000}$ part of that of the globe. The atmosphere is spread over the earth in concentric layers without any regard to the inequalities of the surface on which it rests. The air in deep mines is the most compressed, but the communication through this mobile fluid is so rapid that storms influence the barometer there as surely as on the surface of the earth. At a mile above the level of the sea the air all over the earth has the same density whether it rests on air or on a mountain. It if were not so the barometer would be no test of elevation.

§ **242.** The shape which this atmosphere assumes is undoubtedly that of a spheroid, probably more flattened than that of the earth, because being more distant from the centre centrifugal force bears a larger proportion to gravity than it does at the surface of the earth. If the earth rotated more rapidly the ocean and the atmosphere would be floated from thesurface. It is only the larger globes, such as Jupiter and Saturn, which by their greater mass can counterbalance the centrifugal force generated by their swift rotation.

Had this atmosphere been differently composed, man could not have retained his present physical constitution. The gases which compose the air are adapted to the lungs of its inhabitant; neither gas can be increased without injuring him. The supply of air also is precisely what the lungs require. The diver who fills his lungs with air compressed beneath the water, finds his lungs irritated, while the extreme pressure on his ears, his head, and the whole surface of his body is extremely painful. No less is the suffering experienced by those who ascend lofty mountains. The air in their bodies being less rarified than that around them causes the most unpleasant sensations in the head and ears. The air which they breathe is too rare to satisfy their lungs. Fishes cannot exist in ponds on lofty mountains, for the air in the water has not sufficient oxygen. For the same reason it is difficult to kindle or maintain a fire. Even sound is feebly transmitted by so thin a medium.

§ **243.** In the atmosphere are diffused watery vapor and many gases. Electricity, magnetism, light and heat are busy there, working changes in organic and inorganic nature. The robe of green with which the earth is made ready for man and animals has existed in the atmosphere in a gaseous state. In this delicate form animals and plants imbibe and condense them within their frames to all those several forms which the Infinite Disposer has appointed to matter.

But the atmosphere is not only useful, it is the abode of all which is most beautiful. We scarcely know which to admire most, the pure transparent depths which almost open to us another world, or the many forms of loveliness assumed by those sky-dwellers the clouds.

The sun heating the water and the damp earth causes the moisture to take a vaporous form and consequently to rise. Thus the earth is surrounded by two atmospheres, one of air, one of watery vapor, either of which might exist independent of the other. This vapor is invisible as long as the air which it permeates is sufficiently heated to allow it to remain a vapor. When, by ascending or by meeting

15

a cold stratum of air, it is chilled, it ceases to be a vapor, it becomes a cloud.

§ **244.** Of the precise way in which clouds are formed we are ignorant. Those who have observed them on mountains say that they consist of minute vesicles so light as to float in the air. Perhaps electricity is concerned in their formation, and still more in their precipitation to earth in the form of hail, rain and snow.

Most clouds are less than four miles above the sea level, and many are far below this, and of course below mountain tops.

Probably clouds never exist at a height greater than ten miles, at which height the density of the air is about an eighth part of what it is at the level of the sea. Vapor rising from the earth is condensed before it reaches this height; for the air becomes rapidly colder as we ascend. Air being no conductor of heat is only warmed by contact with the heated earth or sea. The heated layer expands and consequently rises, and the one which takes its place is heated and again rises. The rays of the sun pass through it without imparting heat as light passes through transparent substances. As the heating of successive layers is a slow process, the air is not warmed to any great distance above the general surface; and this is one cause of the cold of mountain tops.

§ **245.** The atmosphere like the sea is never tranquil. Its extreme lightness, and its expansion by heat and consequent change of weight, prevent its ever remaining in a state of rest. This state of agitation not only equalizes the temperature and the moisture, but conduces to the health and vigor of vegetable and animal life. It is only when we feel the breeze that we realize how great a restorer constantly envelopes us. Waves and tides on a larger scale than those of ocean may and probably do exist at the surface of our atmosphere; indeed recent experiments have proved this and have even shown a difference between the influence of the moon when in apogee and in perigee. There must also be an aerial as well as an oceanic equatorial current. As the friction of the particles of water on each other prevents their rotating with as much velocity as the mass of the earth, and thus creates a westward tendency or current particularly in the parts of the earth's surface which have the most rapid motion, thus also, but in a much less degree does the friction of the particles of air cause them to lag behind, and to form a current from east to west. Within the atmosphere we find steady currents and unsteady motions, raised by the unequal heating of the earth's surface. To both of these we give the name of winds. These winds are cold or warm, moist or dry, according to the temperature of the country or ocean from which they blow. The most important currents in our atmosphere are the trade winds, whose regular return enables our navigators to travel with speed and safety over the central portion of the globe. They prevail within the latitudes of 30° N. and 30° S.

As the zone which extends 23° each side of the equator is more nearly in the sun's plane than the other parts of the earth, this zone is more heated than the rest. The air above it is also heated, expands and rises. Air being elastic presses in on all sides to fill up the vacuum. The cold and heavy air from the poles rushes towards the tropics, and the heated air, in an upper current, flows toward the poles, cooling gradually as it passes. We might therefore expect on the earth's surface a steady south wind in the northern hemisphere, and in the southern a steady north wind. And this would be the case but for the greater velocity of the equatorial portions of the atmosphere.

§ **246.** The air appears to be at rest because it partakes of the velocity of motion of that part of the earth's surface on which it rests. Each portion therefore has a velocity of rotation proportioned to that of the circle of latitude to which it corresponds. That near the poles has a very small velocity. Therefore as it rushes toward the equator its speed is insufficient to keep up with the rotation of the earth and the superincumbent air. Thus the currents which otherwise would be north and south, lag, and hang back from east to west in the direction opposite to the earth's rotation, and become north-east and south-east winds. As they approach the equator, the earth, by attraction and by its friction, communicates to them greater velocity of rotation, so that near the equator they become more nearly north and south. Here these opposite currents meeting destroy one another, one or other prevailing locally owing to the distribution of land and water near the equator. As the earth is of nearly the same size for a degree or more each side of the equator, we should expect these currents to be much weakened before they reach the equator. Accordingly here we find a region comparatively calm and free from any steady easterly wind. Ships often are becalmed and wait for weeks to cross the line.

As the heated equatorial air flows over towards the poles, it retains a greater velocity than that of the parts over which it passes. Hence it gains on the earth in an easterly direction. As it gradually sinks to the surface it causes a south-west wind in the northern hemisphere, and a north-west one south of the equator. This is the origin of the westerly winds so prevalent on the Atlantic as usually to make the passage from Europe to America shorter than that in the contrary direction.

If a portion of the air moving slowly were suddenly transferred to that which has a more rapid motion, or the reverse, a most violent shock or hurricane would be the result. Possibly this may be one of the causes of hurricanes.

§ **247.** Beside the trade winds there are other steady local winds, evidently owing to the position of the sun. Some of these, as the monsoons, are caused by the alternate heating of Asia and Africa. A north-easterly wind, probably a portion of the northern current moving towards the equator prevails in summer all over Europe.

Other winds arise from purely local circumstances. Thus the east winds so prevalent on the eastern coast of America in the spring arise from the excess of the heat of the continent over that of the sea constantly cooled by icebergs and polar currents. The west and north-west winds which prevail at other seasons are attributed to the gulf stream, which by heating the air above it draws down cold air from the neighboring parts.

Wherever, as on a desert or a rocky plain, the nature of the soil is such as to become more heated than the surrounding surface, the air is heated above it, and cold air rushes in all round. Columns of air once set in motion rotate rapidly, bearing all things along with them. Their path may be traced by the destruction they leave behind, trees and houses lying so as to show that the rotation not the onward motion of the column has done the mischief. These tornadoes are mentioned here because something like them is observed in the sun's atmosphere. Tornadoes and hurricanes are more common in the tropics, as we might expect. The air having here so rapid a proper motion, a greater shock would be caused by any obstacle thrown in its way. We do not know enough of the arctic regions to pronounce with certainty, but they seem to be the regions of silence and comparative quiet. The temperate zones have been called the battle-ground of the winds. Winds of all directions certainly prevail here, but if their battles are numerous, they are at least not very severe, mere skirmishes compared with the hurricanes of the tropics.

The velocity of wind varies from mere nothing to a hundred miles an hour; and when moving its fastest it has a force of forty-nine pounds on the square foot.

The air at the equator rotates at the rate of 1,000 miles an hour, but as we partake its motion we are perfectly insensible to it.

§ **248.** Since we live at the bottom of the atmosphere and are enveloped in it, it is evident that whatever reaches us from other bodies may undergo changes which we have no means of discovering. The heat which the sun sends us may be divested of some of its most lively properties. Light, whether from the sun or the stars, may have lost some of its constituent parts, or may have altered in color. Perhaps but for this blessed canopy the excess of light and heat we should receive from the sun would be unendurable. Perhaps by some modification of the absorbing and retaining powers of their atmospheres the remotest and the nearest planets, nay even the sun himself, may become inhabitable. Air and water are among the most transparent bodies which we know, yet these when interposed in sufficient quantity absorb great quantities of light. On the summits of the highest mountains, where light passes through a much less extent of air than on a plain, a greater multitude of stars are visible, and through great depths of water objects become almost invisible.

Since air though rare consists of material atoms we cannot be surprised at its effect on light and heat. It has weight, and apparently a blue color, and to this the hue of distant hills is owing. This color of the air must affect the color of all light transmitted through it. We therefore do not any more see the sun of its true color than the diver in the depths of the sea, who beholds it at noon-day of a red color. To him it has undergone two changes in passing through two media; to us, as far as we know, it has undergone but one. Although it has been conjectured that the dark lines and spaces in the solar spectrum may be left by rays of light quenched at his surface. Perhaps in the atmospheres of the other planets blue rays may be absorbed and red or yellow transmitted, and thus another source of variety may be introduced into the system.

§ **249.** Of the amount of all these changes and losses we must continue ignorant. We can never know the quantity of heat or light sent out by the sun, nor the change of color light suffers before it reaches us. But we know from experiments on transparent media some of the laws of these changes, and we can estimate the quantities received from him at different altitudes, or at the same time by two observers with quantities of air interposed, as at the summit and base of the same mountain, and can ascertain whether these estimates agree with the laws we have assigned to transparent media.

Two laws of absorption are well established; first, that light is lost in proportion to the obliquity of the incidence; second, that the loss is proportioned to the density of the medium and the amount of it traversed. The first law may be verified in the nearest pond; vertical rays have more penetrating power than those which fall obliquely; they seem to meet with resistance proportioned to the obliquity of their incidence, and more are therefore quenched or absorbed on their passage. If light is a motion of an ether which penetrates all bodies, we can easily conceive that this motion should be brought to rest more speedily by an oblique than by a direct passage through a medium partially transparent. Now the rays of the vertical sun enter the atmosphere at right angles to its surface, and they enter more and more obliquely as the sun approaches the horizon. From this cause then not so many rays of the declining as of the noon-day sun penetrate our atmosphere.

§ **250.** Let us consider in what direction the greatest quantity of air would be interposed between us and a heavenly body. If we draw a globe surrounded by an atmosphere, and from our position on it draw several lines, one toward the zenith, one at an angle of 45°, and one horizontal, we shall find that the last passes through the lower and denser layers for a much greater distance than either of the others. The vertical line passes through the shortest extent of atmosphere; and each line passes through the air and through the dense layers for a greater distance, in proportion as it approaches the horizon. These lower layers are not only more dense in themselves, but they are usually loaded with vapors and often with clouds. The absorbing power of vapor is seen in the clear crimson and orange light diffused by the setting sun. The vapors then suspended in the air absorb all the other rays, allowing passage to these alone. The whole amount of light and heat absorbed by these vapors must be very great since it enables us to bear the sight of the sun. Hence less light is received toward the horizon than near the zenith because there is actually a larger body of air and vapors to be penetrated near the horizon.

§ **251.** As the heat-giving rays apparently proceed from the same cause and are propagated in the same way as the light-giving rays, and as observation shows the same results, we may conclude that they are subject to the same laws, and we may state the general law, that the absorption of heat and light by the atmosphere increases with the obliquity of the incidence.

Of ten thousand rays falling on its surface, 8,123 arrive at a given point of the earth if they fall perpendiculaly; 7,024 arrive, if the angle of direction be 50°; 2,831 if it be 70° ; and only five rays will arrive through a horizontal stratum. Between one quarter and one fifth of the sun's light is lost in passing through a vertical plane. In passing through a horizontal stratum the light is diminished 1,300 times, and this enables us to look at the setting sun without being dazzled. A haze increases the loss to one third.

On the summits of high mountains, where but little air intervenes between the observer and the heavens, a multitude of small stars may be seen which are invisible from the plain. Within 10° of the horizon small stars become invisible, because the light from them cannot penetrate so dense a medium.

§ **252.** We may form some idea of the quantity of light and heat sent from the sun, by considering that they are transmitted through space in all directions, and incessantly ; that comparatively few of them fall on any globe or object that is known to us, and that if the sun were surrounded, at the distance of Uranus, by a vast hollow globe, it could brighten the walls of that with as much ease as it now gilds the few planets which wander in depths of space. When we reflect on the vast quantity of heat and light received by our earth, which is but a pin's point in the heavens, and on the brightness which the planets and the moon owe to the rays which they receive, we can scarcely conceive of the immense number of rays continually radiated and lost in space.

Various attempts have been made to estimate the light which we receive from the sun. Its direct light has been estimated as equal to 5,563 wax candles of moderate size placed at the distance of one foot from the object; that of the moon is only equal to the light of one candle at the distance of twelve feet. Consequently the light of the sun is more than 300,000 times as great as that of the moon, and that of the moon is too small to afford heat. It would require 90,000 moons, enough to fill the whole of our visible sky, to afford us light equal to that we have in a cloudy day when the sun does not shine out. All the light we have in a cloudy day is by reflection from the clouds, and the light of the moon is only a reflection of that of the sun; therefore it would take 90,000 of them to give us the same light. Compared with Sirius, one of the nearest and largest of the fixed stars, the sun's light is twenty millions of millions of times as great as that of Sirius to us. But Sirius placed where the sun is would appear 3.7 times as large as the sun, and would send out 13.8 times as much light.

§ **253.** We will now consider whether the rays which enter our atmosphere are bent from their true path. This is a more important inquiry, in an astronomical point of view, than their diminution. If rays are bent, we can never see any heavenly body in its true place, unless it is in the zenith. And since to us, in latitude 42°, the sun, moon and planets, are never within 12° of the zenith, it follow that we never see any member of our solar system in its true place, if refraction actually takes place.

As we live at the bottom of the denser medium through which rays pass, it is impossible for us to see their refraction as we can when a ray or a stick passes from air into water. We can only find out by experiments with transparent media what are the laws of refraction, and then ascertain whether the phenomena observed in the heavens agree with those laws.

§ **254.** We find on earth that whenever a ray passes obliquely from a thin to a denser medium it is bent towards a perpendicular to that point of the surface which it touches. It is more bent in proportion to the density of the medium it enters, and also in proportion to the obliquity of incidence. Refraction takes place only at surfaces. At whatever angle a ray enters a medium, at the same angle it proceeds through that medium; therefore refraction does not, like absorption, increase with the thickness of the layer or the extent traversed.

A ray falling perpendicularly on a medium is not refracted, but passes on in its original direction. For oblique rays the sines of the angles of incidence and refraction, (the angles which the ray before and after meeting the surface makes with a perpendicular to that surface at the point where the ray meets it,) bear always, whatever be the amount of the angle of incidence, the same proportion to each other so long as the medium is the same.

If the ray passes through several media bounded by parallel plane surfaces, the effect produced by refraction is the same in amount as if it had originally fallen on the last of these media at the same angle as that at which it fell on the first.

Let us see if in our atmosphere we find the conditions of refraction; and also whether we observe its consequences in the altered position of the heavenly bodies.

§ **255.** Though the outer parts of our atmosphere have been described as exceedingly thin, the surrounding ether is much more rare, or it would affect the motions of the planets. Light passes from the ether to the atmosphere, from a thin to a denser medium; therefore it is refracted, except when it falls perpendicularly; therefore we see all heavenly bodies, except those in the zenith, higher than in their true place. For as we have no knowledge how much rays may have been bent since they left bodies, we refer all bodies to that direction from which light meets our eyes.

If the air were of the same density throughout, rays would be bent once at the surface of the atmosphere, and would then reach the earth without again changing their angle. But as the strata of the air change continually in density as they lie nearer the earth, as they are infinite in number and infinitely thin, the refracted ray describes a curved path. And since in similar media refraction is great in proportion to the density of the compared layers, and the density of the lower layers increases so rapidly, this curve becomes rapidly steeper as it approaches the earth. But we at the bottom of the aerial ocean know nothing of this curvature of the ray; we refer the object to the end of a tangent to the curve at the place where it enters our eye, and of course see it more elevated above our horizon than it really is. If the air were of equal density, the refracted ray would be the straight line joining the ends of its present curve.

§ **256.** The amount of refraction is slightly less on the convex surface of the earth than it would be on a plane surface, for the perpendiculars drawn from each point of the surface of the atmosphere are no longer parallel to the zenith line, and thus the angle of refraction diminishes.

Since refraction increases with the obliquity of the incident rays, it must increase very rapidly near the horizon. Its amount is also much increased by the commotions of the atmosphere and by the moisture floating in it. Strata of different density laid one upon another also have great refractive power. Since the fluctuations of heat, winds and clouds, never extend more than ten miles, and usually take place at a much less height, it follows that refraction is much greater in these lower regions. And as horizontal rays pass through many more of these clouds and vapors, it is not surprizing that refraction increases rapidly as we approach the horizon.

The amount of refraction may be found by observing the greatest and least altitudes of some circumpolar star which passes at or near the zenith. Then knowing the latitude of the place, the distance of the star from the pole at each observation will also be known; as the star is not influenced by refraction in the zenith, the differences of these distances will be the refraction at the least altitude. The influence of refraction in every part of its course, except when in or very near the zenith, is shown by the star's describing a flattened curve instead of a circle. The variation is however too slight to be perceived by the naked eye; it differs at each instant.

§ **257.** The law which refraction follows has been determined, but so many accidental causes modify its amount that observation is found to be a better guide than theory in constructing the tables which are indispensable to correct every altitude of a heavenly body. In the zenith the refraction is nothing; at 45° it is about 1′, a quantity scarcely perceptible by the naked eye; in the visible horizon it is 33′, which is rather more than the greatest apparent diameter of the sun and of the moon. Thus we see the sun's whole disc at rising and setting when it is actually below the horizon. And in some parts of the year it rises five minutes earlier in the morning, and sets five minutes later in the evening, than if there were no refraction. The mean time added from this cause is three minutes.

Since refraction takes place in our atmosphere, it of

course affects all the heavenly bodies however distant. In 1750 a singular consequence of refraction was observed at Paris. The moon suffered eclipse in the shadow of the earth, although both sun and moon were above the horizon, one in the west, the other in the east. The sun was in reality a little above the horizon, the moon a little below this plane, but it was raised and made visible by refraction. If the sun had been a little below and the moon a little above, the same phenomenon might have taken place.

§ **258.** Refraction was first suspected from the different apparent heights of the same star at different seasons, and from the different distances of the same star from the polar star, according as it was more or less near the zenith. This was at first attributed to accidental vapors, and it was not suspected that there was always some refraction. It is now known that refraction takes place every where except at the zenith, although the amount of it at the same height is varied by many circumstances. By measuring the horizontal refraction of the upper limb of the sun when it first appeared in the horizon, and then its lower limb, it has been found that even while the sun was rising, refraction had diminished 25″.

Even terrestrial objects are influenced by refraction. The altitude of a hill is sensibly greater on a cloudy, dull day, when the air is thick and heavy, than on a clear day. It is also greater before sunrise than at the noon of a bright day. Not only the barometer but the thermometer indicates by its changes a change of refraction—increased cold being accompanied by increased refraction. From the increase of refraction in night and in winter, it has been supposed that refractions are proportionately greater toward the poles, and serve to shorten a little the gloomy polar night.

§ **259.** Refraction has a very singular effect on the form and proportions of bodies in the horizon. It has the same upon the form of bodies under water seen from the air. A ring plunged in water looks flattened, because the lower part is more uplifted than the upper. Thus the sun, which always appears round when in the zenith or at an altitude of many degrees, as it approaches the horizon becomes flattened and apparently oval, because the refraction of the lower limb being greater than that of the upper, the vertical diameter is diminished. The horizontal diameter is very slightly diminished, because the vertical lines in which refraction takes place are vertical circles. The convergence is however extremely small. For suppose the diameter of the sun to be 32′, and the lower limb to touch the horizon, then mean refraction at that limb would be 33′, but the altitude of the upper limb being 32′, its refraction is only 28′ 6″. The difference between these refractions is 4′ 54″, the quantity by which the vertical diameter appears shorter than that parallel to the horizon. When a body is not very near the horizon refraction diminishes very nearly uniformly.

Although the lower limb of the sun and moon appear expanded when in the horizon, the whole disc is smaller. Their splendor is diminished, and the cause of this is so frequently increased distance, that we are apt to suppose it the only cause, and to refer dim and indistinct objects to a great distance. Painters expect this reasoning from us; they use indistinct outlines and faint colors for distant objects, bold outlines and strong colors for near objects.

After we have referred a dim object to a greater distance than the true one, by another process of reason we judge it to be larger than it really is, because at so great a distance it appears of a given size.

We refer all bodies near the horizon to a greater distance than we do when they are overhead, because they are all thus dimmed, and we extend this illusion to the form of the arch itself, and believe that we see it more extended near the horizon than overhead. Whereas in fact we see vertically to a greater distance than we do horizontally, for a small star which is visible near the zenith, is invisible near the horizon. So that our vision does not extend to quite a concave hemisphere, it is narrowed near the horizon. The arch does not at all times appear of the same elevation. It is higher in clear than in dull weather.

§ **260.** Perhaps one reason that a horizontal section of the arch appears larger is that we have more means of measuring it. As we look toward the horizon we see stars separated from one another by space, extending in a long series, or we see an alternation of clouds and sky, and beneath, a landscape which we know extends to a great distance. But if we look upwards we have no line of stars, no marks to inform us whether we are gazing miles, or hundreds, or millions of miles, into the heavens.

In the same way our judgment of the size of the moon in the horizon is affected by the vicinity of objects so much smaller as terrestrial objects must be.

Actual measurement proves that when they are in the horizon the sun subtends the same, and the moon a much smaller angle, than when at a greater altitude. Any one may satisfy himself of this by rolling a piece of paper into the form of a tube, making the opening the size of the moon when in the horizon. Tie a thread round it to keep it of the same size, and when the moon comes on the meridian and appears much smaller to the eye, look at her through the tube, and she will appear larger than at her rising. She will appear larger because she is nearer to the observer; of course her actual size cannot vary, for when she is in the zenith of one observer, she is in the horizon of another. Suppose her in the zenith of a person 90° distant from us, and in our horizon. It is evident her distance from us is greater than her distance from the other observer. If her distance is greater she subtends a smaller angle and therefore appears smaller.

Another phenomenon of a somewhat similar nature, is the apparent enlargement of the bright part of the moon when both the bright and dark parts are visible. The unequal impressions made upon the retina of the eye by the bright and the feebly illuminated regions give rise to this illusion.

§ **261.** Refraction likewise occasions twinkling or undulations in the light of the fixed stars. The atmosphere being very easily expanded by heat and condensed by cold, is always more or less agitated. The layers of molecules which compose it experience momentary condensations and dilatations, which cause the direction of the luminous rays to vary incessantly by the difference of refraction which they occasion. Vapors and layers of air of different densities are also drifted rapidly along, so that the refractive power of the medium varies continually. These effects are almost always perceptible in our country, because the air here is seldom serene ; they are less so in countries where the sky is more pure. They occur particularly when the weather is changing. For this reason the fixed stars, whose apparent diameter is very small, appear to us agitated by a sort of trembling. This happens especially just before rain when it succeeds a long dryness. The twinkling of the stars is then so remarkable that it becomes a sign for sailors.

At such times if a star is observed with a delicate telescope, when the star is placed under the thread, it will oscillate so as to appear successively on each side the thread. These motions succeed one another with such rapidity that the visible diameter of the star appears to exceed the thickness of the thread. These motions are sometimes so great that it is impossible to observe.

Very marked agitations, produced by the same cause, may be observed in the shadows of towers, and in the image of the sun projected on the ground by an opening made in the dome of a lofty building.

The rapid motion of the air caused by dilatation may be seen above stoves, or above fields and the roofs of houses when much heated after a long drought. Any object seen through this rapidly changing medium appears distorted and trembling.

We are apt to suppose that the fixed stars look larger than they really do, owing to the false glare occasioned by this trembling. An ordinary telescope magnifies them still more than the changeable air. The imperfections of the glass give the stars spurious discs. But on observing these same stars with telescopes of a much higher power they appear as mere points.

The planets twinkle much less than the fixed stars ; their discs are so much larger that they cannot be displaced totally. They experience on the edges little undulations, while the stars which seem but brilliant points are continually displaced. This displacement produces twinkling.

§ **262.** Refraction is not the only change light experiences in our atmosphere. Reflection does not influence the distant luminaries, but acts at all hours and on every body within our atmosphere.

It is not easy to separate the effects of refraction from those of reflection. Twilight, for instance, is the effect of reflection following refraction; reflection prolongs it more than refraction, but to refraction and absorption we owe the infinite variety of the morning and evening sky. Without their influence there would be a sudden transition from splendor to darkness, from day to perfect night. As refraction brings up the sun's disc when actually below the horizon, so it afterward brings his rays up higher, and makes them visible longer than they would otherwise be.

Reflection causes the rays from a body below the horizon after rising above the horizon and striking against the vapors and clouds, and perhaps the atoms of air, to be sent downward to the earth. If the sky is clear these rays are reflected to us of a pure yellow light. If it is loaded with clouds and vapors at different heights, different colored rays struggle through these, and we have a sky varying in color every instant as the rays strike the clouds more or less obliquely. The red rays have most momentum and therefore pass through a misty sky where no others would. The other colored rays are absorbed. The sun's rays rise sufficiently high in our atmosphere to be reflected until he is 18° below the horizon.

§ **263.** The usual duration of twilight in the temperate zone is an hour and a fifth long. Duration of twilight is increased even more than we should expect on high mountains. De Saussure passed several nights on the high Alps, and saw the whole horizon surrounded with pale but distinct light which lasted from sunset to sunrise, although the sun must in the middle of the night have been 45° below the horizon. This reflection did not come from the layers of air where the observers stood, for on such heights they are so near that their reflection is very feeble, but from the thick and deep mass of air which borders the horizon on all sides.

Analogous phenomena are sometimes seen during an

eclipse of the moon. Her disc is not in different eclipses always of the same color. It has been supposed that when that portion of the sun's atmosphere which is so situated as to reflect the sun's rays upon the moon is laden with vapor, it gives to the moon the peculiar light and color which are sometimes observed.

§ **264**. Reflected light is not seen only in twilight. Almost all the light which falls upon our eyes has been again and again reflected. The light which comes from a bright luminous body is too brilliant to be agreeable; it is painful to the eye. If it fell upon bodies and were only once reflected to the eye, we should see on every object a round brilliant image of the sun, such as is reflected from polished steel and from water. The moon would send us only a reduced image of the sun. Every body in the direct rays of the sun would be painfully brilliant, and all other bodies would be in the deepest darkness; every room into which the sun was not shining would be as dark as in the night. But most objects which the sun shines on are too rough to reflect his image; they break his rays into innumerable smaller rays, and the greater or less brightness of these and the angle at which they touch our eyes, teach us the form of the body. In very distant bodies we lose the difference of shade and the form consequently. Thus the sun and the moon from their great distance show a flat disc. Besides we see no bodies by the direct light of the sun only, but also by an infinity of cross lights, which coming to us from all parts of the body show us its whole shape. These cross lights on the surface of the earth arise partly from reflection from large objects, but more from reflection from vapors and small particles floating in the air, and perhaps also from the particles of air themselves. The power of very small particles to reflect light is shown by the path of a sunbeam across the room, or across a moist atmosphere, when the sun is improperly said to draw up water. Particles of dust and vapor, before invisible, become perfectly luminous, that is break and reflect the rays in all directions, so that they are themselves seen. Undoubtedly much smaller particles have this power, and by their unseen action produce the soft generally diffused light of day. The rays are broken and sent as messengers in all directions, crossing and recrossing and wrapping us in a perfect web of light, till we almost forget that the little disc which our two hands can shut out from our sight is the source of it all.

§ **265.** The indirect light which is reflected from the sky is often very considerable. In Edinburg it amounts perhaps to 30° or 40° of the photometer in summer, and 10° or 15° in winter. This secondary light is most powerful when the sky is overspread with thin fleecy clouds. It is feeblest in two very different conditions,—either when the sun's rays are obstructed by thick clouds, or when the atmosphere is quite clear and of a pure azure tint. In higher regions, the direct rays of the sun, not being impaired by a long passage through the atmosphere, are more vigorous than at the surface of the earth. But the diffuse, indirect light of the sky, being reflected from a rarer mass of air, unstained by vapors, is proportionately feeble. The silvery hue of the sky changes to a dark hue, slightly tinged with blue in the day time, and at night serving as a transparent black ground for the multitude of stars. As this feeble diffused light does not interfere with vision, large stars and planets are visible from the shade even in the day time.

Reflection from the ground and other opaque objects makes no inconsiderable addition to the amount of light. From a sandy beach, the reflected equals one third of the incident light. From a wide surface of snow, it amounts to five sixths of the direct light; the numerous facets of the bright snowy flakes, which are presented in every possible position, detaining only one sixth of the incident rays, and scattering the rest in all directions.

The laws and facts thus far studied concern the globe as a whole. Before entering on those which take place in portions of its surface, we will see what effect the position of an observer has on the appearance of the heavenly bodies.

CHAPTER XII.

PHENOMENA WHICH DIFFER IN DIFFERENT PARTS OF THE EARTH.

Day and Night. Circle of Illumination. Twilight. The Seasons. Curve traced by the Sun's combined daily and yearly Motions. Portion of the Heavens visible in different Latitudes. Length of Day and Night. Equinoxes. Effect of Twilight. Amount of Light and Heat received in a given place. Equality of the distribution of Heat in the Northern and Southern Hemispheres. Difference in their respective Seasons.

§ **266.** Before we enter on those celestial phenomena which appear different when viewed from different parts of the earth, we must inquire what effect our position on the surface of the earth has on external phenomena. The utmost distance by which two observers on the earth's surface can be separated is 8,000 miles; and 8,000 miles is an appreciable distance to bodies no more distant than the sun, moon and planets. It is sufficiently large to be distinguished from them, and causes them to change their places as viewed from the earth. This displacement of a body from its true place, as seen from the earth's centre, is called parallax. Owing to it we never see a heavenly body in its true place unless when the line joining it and our eye passes through the centre of the earth. The displacement is greatest when the body is in the horizon of the observer, because the observer is then distant from the line joining the body to the centre of the earth by the greatest possible amount, the earth's radius. It is then called the horizontal parallax. Its amount may be observed for the sun, moon and planets; it is greatest for near bodies, less for those more distant, and inappreciable for the fixed stars. It is less for each body according to its height above the horizon; it is only in the case of a body in the zenith that it becomes nothing, and in high latitudes this can never take place with any member of the solar system. It always acts in a vertical circle, and always depresses the body, thus partially counteracting the effects of refraction.

§ **267.** The laws we have hitherto investigated have concerned the globe as a whole. We will now study some phenomena which are unlike in different parts of the globe, the variations of seasons and in the length of days. An astronomical day includes twenty-four hours, a natural day may be of any length between nothing and six months. The various modes of reckoning days and years will be mentioned hereafter; at present we have only to account for the length of days at different latitudes, and in the same latitudes at different parts of the year.

To do this we must return to the position of our earth and its two motions; and we must imagine the equatorial and the ecliptic to be marked out on the concave sphere of the heavens; and must remember that while the earth by its rotation causes the sun to appear to move from east to west in the equator or parallel to it, by its motion of revolution it causes the sun to appear to move in the ecliptic from west to east.

One half of this earth's surface is always illuminated, the other half in darkness. This illuminated hemisphere has its edges bounded by a great circle called the circle of illumination. In the spring and autumnal equinoxes it is bounded by a meridian, called the solstitial colure. At no other season of the year is it bounded by a meridian. The illuminated hemisphere extends 90° in every direction from the point to which the sun is at each moment vertical. This point is always in the ecliptic. The illuminated hemisphere therefore may extend 23° beyond either pole, or fall 23° short of it. Since rotation never allows the same spot to remain beneath the sun's vertical rays a moment, the earth each moment turns up a different hemisphere to be illuminated.

§ **268.** Let us suppose a concave hemisphere of light to be fastened directly between the earth and the sun in the plane of the earth's revolution, and to move round in the course of the year so as to represent the light falling from the sun. Then let us imagine the earth performing her rotation in a plane 23° inclined to this hemisphere of light, and we shall see that the space for 23° round the pole will rotate sometimes entirely in light, sometimes entirely in darkness.

Let us begin at the vernal equinox; at this time the illuminated hemisphere extends to both poles, and just one half of the northern and one half of the southern hemispheres are illumined. The next day at noon, in the same place, the centre of illumination lies north of the equator a few minutes of a degree, therefore the illumination extends a few minutes over the north pole, and falls a few minutes short of the south pole. Its edges no longer coincide with terrestrial meridians; more of the northern than the southern hemisphere of the earth is included in it. After another rotation the centre of illumination is farther north. And this continues for one quarter of a year. At this time the illumination extends 23° beyond the north pole; it has covered the north pole for one quarter of the year, and will continue to do so through the next quarter. For the next three months the central point advances in the ecliptic, always approaching nearer the equator. At the autumnal equinox the north pole is left out of the illumined circle not to enter it for six months. For the next six months the illumination creeps slowly over the south pole, and then retires from it, till the vernal equinox again equalizes the northern and southern hemispheres. In its daily revolution the earth turns up 15° of the equator every hour. Therefore the illumination extends 15° more to the westward every hour. Thus in the course of twenty-four hours every meridian has entered the circle of illumination, been under the sun, and left the circle of illumination.

§ **269.** We have said that the sun illumines a hemisphere, or 90° in every direction from the vertical point. It in fact illumines a little more than this, for the rays which proceed from the outer portions of the sun, illumine a small portion of the otherwise dark hemisphere of the earth. For if the shadow cast were cylindrical, one hemisphere would be illumined, and the rays from the edge of the sun's disc would graze the edge of the earth's disc. In this case the sun would be precisely the size of the earth. And it would make no difference whether the distance between them were great or small.

But let us suppose the sun smaller than the earth. It

would then illumine less than half the earth's surface, because the rays from the sun's disc, diverging, would meet the earth's surface before they reached the large circle running at right angles to the sun's direction, which divides it into hemispheres. And the ring between this great circle and the circle formed by the sun's rays would be in darkness. In this case if the sun were near, the ring would be broader than if he were more distant.

But since the sun is larger than the earth, the outer rays which strike the earth converge, and not only one hemisphere but a ring adjacent to it must be illumined. If the sun were of the same size he now is and nearer to the earth, this illumined ring would be broader than it now is. But as he is 95,000,000 of miles distant, its breadth is only equal in minutes of a degree to 15′, about one half the sun's apparent diameter. Owing therefore to the superior size of the sun, a ring-shaped surface, 15′ in breadth encompasses the illumined hemisphere and is added to it.

§ **270.** If the earth had no atmosphere this would be the only addition to the illuminated hemisphere. But we have seen that refraction suffered in the atmosphere makes bodies visible 33′ before they are above the horizon. Therefore refraction extends still further the illumination. It adds a fringe of light 33′ broad.

That portion of the earth's surface which lies in twilight cannot properly be said to belong to the illumined or to the dark hemisphere. Twilight is usually reckoned to last until small stars become visible, and this is usually when the sun is 18° below the horizon.

If then we are considering how large a portion of the earth's surface is fully illumined, we should say it extended $90°+(15'+33'=49')$ from the place to which the sun is at each moment vertical. For since the band 15′ is actually illumined refraction acts beyond it. Beyond both of these bands is the ring 18° wide which is faintly illumined. Thus we have $90°—18°—49'=71°\ 11'$ for the radius of the convex surface which remains in darkness.

Even here, however, we cannot say with certainty that some light is not received from the sun. If so it has undergone so many reflections in the atmosphere as to be extremely faint. Probably it is only toward the centre of the dark portion that the darkness is as great as if the sun were blotted from the heavens.

§ **271.** Let us now consider what points of the earth's surface will enter the centre of illumination in the course of each day and of the year. Or what is more easily conceived of, let us consider to what points of the earth's surface the sun will be vertical, beginning with the vernal equinox, on about the 21st of March. Let the earth at noon on this day present to the direct rays of the sun that point which is common to the planes of her rotation and her revolution, and thus the sun will at noon be vertical to this point.

But this point is beneath the sun only the smallest fraction of time. The incessant rotation of the earth withdraws the point from the vertical sun, substituting first one and then another and another point, each west of the preceding, until in twenty-four hours it has been noon to every point in a curve not coinciding precisely with the equator, but beginning in it, surrounding the earth, and gradually rising, never so much as 24′, and seldom nearly so much, north of it. The course which the vertical sun marks out on the earth the next day is a continuation of this curve, almost parallel to the former, but ending a little farther north. The whole course of the sun is a spiral, made up of these curves, lying one above and slightly inclined to each other, like threads wound skilfully on a ball. The threads should not, as we shall explain hereafter, all be equally inclined one to another.

§ **272.** For three months this spiral winds gradually northwards, the path one day being almost parallel to that of the preceding, and almost coinciding with it. At this time the north pole is inclined toward the sun, and the most northerly part which ever comes beneath the vertical sun is now exposed to it. As the earth moves on with her pole fixed, the sun will no longer be vertical to this northerly point, but to one nearer and again nearer to the equator. For several days the vertical sun describes circles nearly parallel to the equator; the sun appears noon after noon in the same spot—it appears to stand still in the heavens. It is called the summer solstice, (from *sol* and *stat*). The reason of this apparent standing still is that the ecliptic and the equator are really more nearly parallel in this part than elsewhere, so that the latter degrees he ascends differ less than any others. And also that if we take the day when the sun described the most northerly circle, and compare that with the one preceding and the one succeeding, the variation during these three days is very much less than the difference in declination between the place of the sun any other three successive noons—as when we run up and then make one step down a hill.

The spiral which we have traced for three months continues for the next three months to wind gradually descending curves round the earth till, in the autumn, it reaches the autumnal equinox. It still descends for three months more till it reaches the southern solstice, and then ascends for three months, having described around the earth 365 curves, almost parallel to one another.

§ **273.** Thus the apparent path of the sun is caused by our two motions. On account of our revolution round him, we think we see him describe a great circle round us once a year, from west to east. And, since this revolution is not in the plane of our rotation, but one half above and one half below it, and $\frac{1}{365}$ of it must be described every day, the sun, in order to pass through it, is obliged to fall nearly one degree short of a circle, and also to move a little farther north or south every day. Since our rotation is always at right angles to the equator, the sun's daily path must be parallel to it, allowing only for his slight change in declination; this change being divided among 24 hours, and being at most 24′, is not perceptible.

By this simple and beautiful arrangement, the point of greatest heat, which if stationary would destroy vegetation and animal life, traverses the globe incessantly, returning to the same spot only once a year, and then, owing to other causes, to a place not precisely the same. While for 90° in every direction from this spot the earth is warmed and illumined.

§ **274.** Let us suppose one spectator at the pole, another at the equator, and a third in some latitude between the two; these are the only important varieties of position which can occur. We will inquire how many and what stars are seen by each observer at night, and in what cases and how the stars seen vary at different seasons of the year. We will learn the apparent motions of the stars and sun to each place, and the length of time they remain visible.

In all these examinations we shall begin with the position of the globe at the vernal equinox, because at that season the planes of rotation and revolution intersect, and the sphere is right for all but the inhabitants at the pole. As soon as an observer is carried out of the illuminated surface, the sun is below his horizon. At that moment, were it not for twilight, he would be in darkness.

§ **275.** To an observer on the equator, in the vernal equinox, the moment the sun is below the horizon twilight begins and lasts till he is 18° below. Were it not for twilight the observer would immediately begin to see the stars, and would see a hemisphere of them extending from pole to pole and as far as to within one degree of the sun. As the earth turns round he would in an hour lose sight on the west of all the stars contained from pole to pole within two hour circles, and he would gain sight of a new strip of the heavens contained on the east within two hour circles and reaching from pole to pole. Every hour of the night makes a strip of the eastern sky visible, and leaves a strip of the western sky out of sight. In the course of the night all the stars of the heavens would be seen were it not for twilight, for at the dawn of a twelve hours' day an observer is 180° from where he was at sunset the night before; and as his horizon in both cases extends over half the celestial sphere, he would in the course of the night see all of it. Twilight makes a strip, extending from 18° east of the sun at night to 18° west of him in the morning, invisible. As the earth moves on in her orbit, the sun is referred to different stars. Different stars are therefore concealed after his setting and before his rising. The invisible portion extends from pole to pole, and is 36° in breadth at the equinoctial. Since the earth advances nearly a degree daily in her orbit, it will require only thirty-seven or thirty-eight days to render this whole strip visible. Every night, however, different stars disappear after the sun. Those which set some time after him one night follow him immediately the next. And those which rise with him one day, the next day rise before him, and he is immediately preceded by stars before unseen. No two nights in the year have the same stars been in the meridian at midnight. Those which passed it at twelve o'clock one night pass before that the next night.

§ **276.** The observer at the equator is, as we shall hereafter show, the only one who, even in the course of a year, sees all the stars. His horizon always stretches from pole to pole. The stars describe paths at right angles to his horizon, and they rise to him always in the same place, though not always in the same time. They appear to move directly from east to west, and all describe in the heavens parallel arcs.

An observer at the poles would have the same hemisphere of stars always visible sweeping round his horizon in parallel circles. Those near the horizon would appear to describe large circles, those near the zenith smaller circles. Over head would be the polar star, to him as to all other spectators, immovable. Revolution never makes more stars visible, but it makes the whole hemisphere invisible during the six months' day while the sun is above the equator. A man at the poles then sees only half the stars which the inhabitant of the equator sees, and these only for one half of the time.

If the sun were always in the equator he would be always in the horizon of an observer at the poles, and thus he might live utterly unconscious of the stars and of all but the brightest planets. He would see the moon as a faint white cloud.

§ **277.** To an observer between the poles and the equator the fixed stars have their nightly courses oblique to the horizon, and more oblique in proportion as the observer is nearer to the pole.

If an observer at the equator has both poles in the horizon, an observer ever so little north of the equator loses sight of the south pole and sees a little beyond the north pole. Let the observer travel 10° north of the equator; then all the stars within 10° of the south pole will sink below his horizon and remain always invisible. All clusters within 10° of the north pole will be in his horizon during their whole course. They describe circles round the pole, being half the time above and half the time below it, and are called circumpolar stars. To such an observer 10° from the pole is the circle of perpetual apparition.

If he advances 20° north of the equator, the circle of perpetual apparition will extend 20° from the pole. However far he advances, the altitude of the elevated pole above his horizon always equals his latitude, and all the stars between the pole and his horizon will be perpetually visible. They will revolve round the pole, keeping their relative position and configuration. The Great Bear will have his feet always turned from the pole, the Little Bear will have his feet toward it. The stars out of the circle of perpetual apparition rise obliquely and describe very large arcs of circles, more than semi-circumferences. As the traveller approaches the pole, they rise with a smaller angle to the horizon, and at last the arcs are almost parallel to his horizon. Those which rise precisely in the east describe a semi-circumference which does not pass through his zenith, and remain in sight precisely twelve hours, and set precisely in the west. The arc described by these stars coincides with the celestial equator. Since the rational horizon is a great circle of the sphere, when it does not coincide with the equator, it must bisect it and be bisected by it. This intersection will take place in the east and west points of the horizon, 90° from the north and south points of the horizon.

§ **278.** The great circle which passes through these east and west points and through the zenith is called the prime vertical. When stars come to the prime vertical they are said to be in the east. A body on the same side of the equator with the beholder rises between the east and north, and comes to the prime vertical after it has risen; a body in the equator rises in the east; one on the opposite side of the equator rises between the east and south, and has passed the prime vertical before it rises. A body on the same side of the equator with the spectator is longer above than below the horizon; one on the equator is as long above as below it; one on the other side of the equator is below longer than above it.

The nearer the observer is to the equator the greater is the number of stars he can see. The circumpolar stars are always the same for a given latitude; their change of place alone marks the different hours of the night. But all the other stars rise and set, new ones constantly appear in the east, describing arcs similar to those of their predecessors, and disappear toward the west. Since the observer at the equator in a few nights sees all the stars, and one at the poles never sees more than half, observers between the equator and the poles see between the whole and the half, and more as they approach the equator.

§ **279.** To two observers in the same longitude, but differing in latitude, the heavens present different aspects at all moments. The stars which are common to both describe circles differently inclined to their horizons, and differently divided by them, and attain different altitudes; and some stars are seen by one and not seen by the other. To observers situated on the same parallel of latitude, and differing only in longitude, the heavens present the same aspects, but at different times. Their visible portions are the same; and the same stars describe circles equally inclined and similarly divided by their horizons, and attain the same altitudes. In the former case there is, in the latter there is not, any thing in the appearance of the heavens, watched through a whole diurnal rotation, which indicates a difference of place in the observer. The only way for an observer in north latitude to increase the number of stars visible is to travel toward the south. As he moves southward, on the same meridian, most of the stars before seen change their places and times of rising and the angles they make with his horizon, and they remain visible longer. Those which rise with the equinoctial, rise in the same time and place as before, but they first make oblique and then right angles with his horizon, and when he has passed the equator will appear to describe arcs inclined toward the north. The stars round the south pole now become perpetually visible ; those within an equal distance from the north pole remain invisible. The change of stars is more noticed as the horizon advances south, because at some seasons of the year and some hours of the night very remarkable stars and constellations appear in the southern hemisphere.

§ **280.** The apparent path of the sun for each of the three observers varies in the same way. If a place is on the equator the sun will always rise at right angles to his horizon. If a place is at the pole, the sun will move always parallel to its horizon. If a place is any where between the tropics and the pole, the sun will rise obliquely to its horizon, and more obliquely as the place is nearer the poles. If it is between the equator and the tropics, it will rise twice a year at right angles, and the rest of the year obliquely. To a place on the equator at the vernal equinox, the sun rises precisely in the east, passes through the zenith, giving twelve hours of daylight, and sets in the west. On the next day, at the same time, it rises five sixths of its own diameter further north, and sets after twelve hours as much further north of west as it rose north of east. The arc described is parallel to the arc described the preceding day, but does not pass through the zenith. The next day it again rises five sixths of its diameter further north, and at the same hour, for the horizon of the man at the equator covers one half the hour circles of the globe, describes another arc nearly parallel to the equator, and sets further north than the day before. This continues for three months, till the sun has risen and set as far north as it ever goes, that is to say, 23° 28′ north latitude. As it approaches the solstice, however, it rises only one sixth of its diameter more north. The arc it describes on our midsummer day to those in the equator begins 23° north of the eastern point, passes 23° north of the zenith, and ends 23° north of the western point of the horizon. It is the lowest arc the sun ever describes to those on the equator ; but like all the others it occupies twelve hours.

If we consider the motion of the observer instead of the apparent motion of the sun, we should describe these changes thus:—In the vernal equinox the inhabitant at the equator is carried through the centre of illumination. Each day he passes farther and farther south of this centre, until, on mid-summer day, he passes 23° south of it, and the day, though of equal length, has less heat than the preceding days.

§ **281.** A man at the pole, on the vernal equinox grazes the northern edge of the illuminated hemisphere. He remains in it, and at noon the next day he has advanced south of the spot where he was the preceding day.

The path of the sun will be parallel to the horizon of an observer at the pole. On the day of the vernal equinox, the sun will appear half visible above his horizon, and will describe a circle around it in twenty-four hours. At the end of these twenty-four hours, it will have risen a little above the horizon. In the next twenty-four hours, it describes another circle, and gains a little in altitude. This continues for three months, during which he is in sight all the time. He then describes circles, gradually descending, till, in three months more, he sinks beneath the horizon, for a six months' absence.

§ **282.** A person in latitude 40° north will, on the vernal equinox, enter the illuminated hemisphere at right angles to its edge; he will describe an arc which passes 40° from the centre of illumination, and will leave the illuminated hemisphere at right angles after a day of twelve hours long. The next day he will enter the illuminated hemisphere, not at right angles, but obliquely toward the south, and in a point farther south, will pass a part of a degree nearer to the centre, and have a day of more than twelve hours. The next day he enters still farther south and more obliquely, passes nearer the centre of illumination, and has a longer day. At the end of three months he has passed within 17° of the illumined centre, and had his longest day. For the succeeding three months the arcs he describes grow less and less oblique to the horizon and shorter and shorter, till at the autumnal equinox he enters the illuminated circle where he did on the vernal equinox, and describes an arc at right angles to the circle of illumination.

For three months after this he enters farther north and less and less deeply into the illumined hemisphere, describing oblique arcs, until he passes $40° + 23° = 63°$ distant from the illumined centre, and has his shortest day.

To such an observer the sun in the vernal equinox would rise precisely in the east, pass within 40° of his zenith, and set in the west. The next day he would rise five sixths of his own diameter north of east, and pass nearer the zenith, and set as much further north. At the solstice, he would rise only one sixth further north. He would then describe his longest arc, and pass within 17° of the zenith. After this his arcs would shorten and decline, till in the winter solstice the shortest arc would not pass within 63° of the zenith.

The more near a place is to the pole the more oblique are the arcs described, and of course the less near they approach the zenith of the place.

A place 10° from the pole enters the illuminated circle when the sun is 10° south of the equator, but passes through only a small portion of it, and is in it but a few hours. Each succeeding day it cuts deeper into the circle, and remains in it longer; and when the sun is 10° north of the equator, it remains in the circle all the twenty-four hours, describing small circles within it, until the sun has been to the tropic and returned to 10° north of the equator. Then it describes large, and continually decreasing arcs of circles, till at last, when the sun is 10° south latitude, it scarcely grazes the illuminated hemisphere. All places within 23° 28′, and not at the pole, describe in this way arcs and circles in the illuminated hemisphere, each with a radius equal to its own distance from the pole.

§ **283.** To a person 10° from the pole, the sun is visible as soon as he is 10° south of the equator. On the first day he merely appears in the horizon for a short time, coming perhaps not half way up the trees. The next day he rises much farther toward the east, and describes a higher arch, and sets much farther toward the west. Each successive day he rises much earlier, and sets much later than the day previous. When he is 10° north of the equator, perpetual day begins at a place 10° south of the pole. He describes in the heavens circles oblique to the horizon. The circles grow higher for three months, and then decline. When the sun is within 10° of the equator, the lower edge of the circle begins to dip below the horizon. Smaller and smaller arcs of oblique circles are described. At length the sun merely shows himself in the horizon ; then his place is marked only by twilight ; and at last he disappears for several weeks.

That spiral of the sun which appeared like successive circles to the observer at the poles, appears like very oblique and large arcs of circles as he travels south. The more obliquely the arcs cut the horizon of any place, the more difference of time is there between two successive risings of the sun, the more rapidly do the days vary. As we approach the tropics the arcs depart more and more from the parallel position they had at the poles, and approach the rectangular position they had to an observer at the equator; that is, they become less and less oblique to the horizon. There is less and less difference in the length of successive days, and less and less difference between the longest and shortest day of each place.

§ **284.** Having learned that places in different latitudes are exposed to the sun for different lengths of time, we are prepared to find the variations in the length of the day in different portions of the globe. Having learned that the same place at different parts of the year remains more or less time exposed to his rays, we are prepared to find the variations in the length of the day for those places where it varies.

In considering the length of the day, we shall take no notice of refraction, twilight, or other causes which influence it.

Since the day of a place at the poles begins when the sun enters the northern hemisphere, and ends when he leaves it, it is six months in length.

The shortest day for a place 5° south of the north pole begins when the sun is 5° south of the equator. It is but a few hours long. The next day the sun remains above the horizon longer. The days rapidly lengthen. When the sun is 5° north of the equator, the longest day to that place begins. It lasts some months, until the sun is again in 5° north latitude. The days then rapidly decrease in length, the last is of the same length as the first, and a night equal to the longest day ensues.

§ **285.** Between the pole and the polar circle the length of the day changes very rapidly, as may be understood by considering how each place rises round the pole. For every degree which illumination advances northward enables very large portions of daily arcs to rise. Of course the nearer a place is to the pole, and the longer the longest day is, the more rapidly the days of each place must change in length. For when the sun crosses the equinox the days and nights are equal all over the world. The longest day for all places in the northern hemisphere falls on the same day of the year; and as those which are nearer the pole have longer days, the increase or diminution of each of their successive days must be greater. Accordingly within the polar circles which are 23° 28′ from the pole, we find the greatest range of variation in the days. At the polar circle there is one day of twenty-four hours, and a night of equal length. Within it the longest days are weeks, and as we approach the poles, months long. Each place, however, not on the pole itself, has some short days.

Between the polar circles and the tropics, the days are, one half of the year, between twelve and twenty-four hours, and the other half they are less than twelve hours long. As the decrease in length is divided among ninety-one days, one quarter of a year, the change from day to day is scarcely perceptible. The longest day for all places in the northern hemisphere is when the sun is in the northern solstice; the shortest day for all such places is when he is in the southern solstice.

At the equator the days are always twelve hours long, for the equator and the circle of illumination being both great circles, must always bisect each other. Twice a year the sun moves in the equator, as all horizons all over the globe bisect his path, and the days and nights are equal all over the globe.

§ **286.** Thus we find that the 4,383 hours of light, and the 4,383 of darkness, which make up the year, are variously distributed in different parts of the globe. At the poles all the hours of light form one period, all those of darkness form another. At the equator the 4,383 hours of light are divided into 365 equal periods, and those of darkness are distributed in the same manner. For all places between the polar circles and the equator, the 4,383 hours of daylight are divided into 365 unequal portions, and for places in the same latitude, north or south, the division is similar. For places between the polar circles and the poles, there are more than one and less than 365 periods of light. The shortest day of each latitude is equal to the shortest night, and the longest day to the longest night. The shortest and the longest day added together equal twenty-four or some multiple of twenty-four hours, and the number of hours of daylight, in a year, equals the number of hours of darkness, for each place on the globe. There is, however, a slight exception to this rigorous distribution of light and darkness, owing to the motion of the earth in her orbit.

Since the earth's orbitual motion is more rapid when the sun is in his southern declination, the days in which he is longer below than above the horizon are not for places north of the equator quite so numerous as those in which he is longer above it, so that the preceding calculation is not strictly correct. The motion is greatest when he is nearest the earth, and least when he is farthest from her. As he is farthest from her nearly at the summer solstice of the northern hemisphere, and nearest to her nearly at the winter solstice, his motion is slower in our summer than in our winter, and the days vary still more slowly in length at the summer than at the winter solstice. For this same reason the sun takes longer in describing that half of his orbit he is in, in our summer, than that he is in, in our winter. In fact, the summer of the northern hemisphere is 7 days, 16 hours, and 50 minutes longer than that of southern.

§ **287.** The whole number of daylight hours near the pole is also increased by the greater amount of refraction in the cold dense atmosphere ; and thus we find that, contrary to what we might expect, the inhabitant of the pole enjoys more of the sun's light than the inhabitant of the equator. There is some evidence that the polar day is prolonged at least a month by refraction. Three Hollanders, who wintered in Nova Zembla, latitude 75°, after three months of continual night, saw the sun rise at noon, a fortnight sooner than they expected ; this could only be owing to refraction. In these cold regions the reflection from the ice and snow is very great ; the aurora borealis appears also with a splendor quite unknown in milder climates ; so that without the moon it is frequently so light all night that fine print can be distinctly read. The moon also rises higher, and is visible longer during the winter than the summer months, except at the equator. At the poles the winter full moon is visible a fortnight at a time, and circles round the horizon like the summer sun.

The following table shows the length of the longest day in each latitude, refraction being allowed for. At the polar circles, those which geographers call hour climates terminate, and month climates commence.

Latitude.	*Longest Day.* Hours.	Minutes.
7° 18′	12	30
15° 36′	13	
23° 8′	13	30
29° 49′	14	
35° 35′	14	30
40° 32′	15	
44° 42′	15	30
48° 1′	16	
53° 46′	17	
57° 44′	18	
60° 39′	19	
62° 4′	20	
65° 10′	22	
65° 54′	24	

Latitude.	*North Latitude.*		*South Latitude.*	
	Continual Day.	Continual Night.	Continual Day.	Continual Night.
	Days.	Days.	Days.	Days.
66° 53′	31	27	30	28
69° 30′	62	58	60	59
73°	93	87	89	88
78° 6′	124	117	120	118
84°	156	148	150	148
90°	186	179	178	177

§ **288.** The duration of twilight has been determined very differently by different observers, and in various parts of the globe. There are many difficulties in determining its length. There is no exact degree of faintness fixed on as its close, though it is generally considered to end when the smallest stars in the west become visible. Perhaps the duration of the morning twilight might be more precisely determined than that of the evening twilight, for the eye, dazzled by the sun, cannot decide when the last ray of light has disappeared. And as the sun between the tropics rushes down almost at right angles with the horizon, he retains his splendor almost to the last, and the dazzled eye is apt to over estimate the succeeding gloom. Whereas, near the poles, the sun descends very obliquely to the horizon, dimmed by the lower layers of the atmosphere for some time before he disappears. The eye thus gradually accustomed to the absence of light is prepared to over estimate the twilight. Perhaps this consideration explains the great diversity of lengths which have been ascribed to twilight, some observers having computed it to last until the sun is 6° or 7°, and others till he is upwards of 20° below the horizon.

§ **289.** If however we allow twilight to last until the sun is 18° below the horizon, its duration in time varies exceedingly to observers in different positions. If we recall what was said of the apparent motions of the sun to an observer on the equator, another at the poles, and a third between the two, we shall understand this.

Let us suppose, round the horizon of each of these observers, a band of 18° added, to represent the extent of twilight. Twilight will last till the sun has left this circle. The sun is at right angles to the horizon of the man at the equator, therefore every degree the sun moves will remove him a degree below the horizon. The horizon of the oblique sphere he will leave obliquely, therefore 18° of his course will not carry him 18° from the horizon, nor consequently out of the twilight circle. The more oblique the sphere is, or the nearer the observer is to the poles, the more degrees the sun must pass through in order to quit the twilight circle.

The shortest possible twilight takes place at the equator at the time of the equinoxes. Its duration is one hour, twelve minutes. The longest possible twilight takes place at the poles. There are there but two twilights in a year; one after the sun has gone below the equator, which continues about 39 days, and another, (which, owing to the present position of the perihelion, lasts rather longer,) before he again crosses the equator. As he is never more than 23° below, the period of total darkness is reduced to about 100 days. The passage from light to darkness is very gradual, and during almost all the twilight there is probably light enough for many purposes.

The longest twilight for places which have the sphere oblique takes place in their mid-summer. In latitude 42° it is two hours, twenty minutes. In the latitude of London, it lasts all of every night from the 22d of May to July 1st.

§ **290.** The variation of seasons is caused by the inclination of the planes of the ecliptic and the equator, or by the non-coïncidence of their poles. This cause operates both directly and indirectly; directly by allowing fewer rays to fall on a country at one season than another, indirectly by influencing the length of the periods during which they fall.

The sun is so distant that his rays may be considered parallel without any want of exactness.

Let us consider how the form of the earth influences the number of rays received on a given portion of its surface. If from the point of the earth which is vertically exposed to them, a plane be supposed tangent to that point, and bounded by a circle the size of the equator, this plane will intercept all the rays which would otherwise have fallen on a hemisphere. They are received on a smaller surface; therefore any measured portion of the plane surface has received more than an equal measured portion of the convex surface. A plane surface exposed at right angles to rain is quickly wet, an oblique one receives less rain for its size, a vertical one may escape without a wetting. If we draw several parallel rays, and cross them at right angles by one straight line, and also by a curved line, we shall find that it requires the curved line to be longer than the straight line to intercept an equal number of rays. Now a sphere falls off in every direction from the vertical point; therefore the number of rays received on a given extent of surface diminishes very rapidly in proportion to the distance from the vertical point. Let us imagine the changes in the obliquity of the rays on any particular spot, from morning until night, owing to the rotation of the earth and the variations in the heat received. Between morning, noon and night, a place would receive the sun's rays at every degree of obliquity which its latitude allowed. In the morning and evening, when the obliquity is very great, the heat received is little; at noon, when the obliquity is lessened, the heat, as we all know, is much greater.

§ **291.** The change in the obliquity of the sun's rays, owing to the obliquity of the ecliptic, does not in the whole year amount to the change in a single day from rotation, but it is sufficient to make perceptible variations in the heat.

As the equator is inclined to the ecliptic 23° 28′, no point more than 23° 28′ distant from the equator can ever have the sun vertical to it. No country within the tropics can ever receive the sun's rays at an obliquity so great as 46° 56′, and twice during one half the year every such country has them vertical. In latitude 42° north we receive his rays at mid-summer at an angle of $42°-23°=19°$; in mid-winter, at an angle of $42°+23°=65°$; in the equinoxes, at an angle of 42°. At the arctic circle the midsummer rays have an incidence of about 43°; the mid-winter one of 90°. At the pole the greatest incidence is $90°-23°=67°$; the least is at the equinox$=90°$.

We have treated the surface of the earth as if it were smooth. The sides of hills receive the sun's rays at a different angle from a plain. The south side of a hill receives the sun's rays less obliquely than the plain, and is therefore warmer.

At the equator and at all places between the tropics the sun is vertical twice a year. Twice a year in such places he casts no shadow. One part of the year the shadows lie toward the north, one part of the year toward the south. At the tropics, he is vertical once a year. North of the tropics, he is never vertical, and shadows fall toward the north. South of the tropics, they always fall toward the south.

At the equator there are two summers, which are at their highest when the sun is in the equinoxes, and two winters, one when the sun rises and sets in the north, one when he rises and sets in the south.

At the tropics the summer is hotter than that of the equator, because the sun is nearly vertical for several days, and the days are more than twelve hours long. The winter, however, is much colder than any season at the equator.

§ **292.** As the longest days at each place must occur when the sun is least oblique to it, the two causes of increased heat act together. Long days give the sun time to act. The earth being a rough, dark, opaque body absorbs heat. All that portion which is exposed to the sun absorbs, all that portion which is turned from it gives out heat. Near the poles heat is absorbed uninterruptedly for almost six months, and there is an exceedingly hot summer; for another six months heat is given out, and there is an extremely cold winter. At the equator, as the days and nights are always equal, this second cause of variety does not operate, and there is very little change of temperature. But in other parts of the globe, the great disparity in the length of days influences greatly the degree to which they are heated. As soon as the day, in any parallel of latitude, begins to be longer than the night, there is a surplus of heat retained through the night. The days continue to increase, the nights to diminish; the surplus is more each day, the amount increases rapidly.

But the day of greatest heat is not the longest day, nor the day of the sun's greatest altitude, any more than the hottest hour of the day is twelve o'clock. The earth continues to receive more heat than it parts with for some time after the summer solstice; this is added to what was before accumulated, and makes the months of July and August hotter than June. In the same way it continues to part with more heat than it receives long after the winter solstice, so that January and February are our coldest months.

§ **293.** We can see what the effect of this second cause of variety without the former would be, by considering the days of a place between the tropics, but not in the equator. Such a place has four days in the year in which the sun attains at noon an equal altitude. Two of these days the sun is north and two south of him. The altitude of the sun then would there be no cause of change. But on the two days when the sun is nearer the tropic he would remain longer above the horizon than the other two. Therefore those two days must be hotter than those in which the sun passes at an equal distance the other side of the zenith.

The amount of heat parted with at each place, equals that received in the course of the year, and thus the mean temperature of each place remains unchanged.

The amounts received in the northern and southern hemispheres are nearly equal. We might suppose, that, as the earth is nearer the sun in the northern winter than in the summer, the northern hemisphere might in the course of the year receive more heat than the southern. But since the earth moves faster in our winter than in our summer, it receives in the course of it no more heat than in our prolonged summer.

§ **294.** The ellipticity of the earth's orbit amounts to one thirtieth of its mean distance; therefore the sun's direct heating power varies one fifteenth; $30^2-29^2=59$; $\frac{59}{900}$=one fifteenth nearly. This would be sufficient to exaggerate the difference of summer and winter in the southern hemisphere, and to moderate it in the northern. But no such effect is produced. For heat diminishes in intensity according to the inverse proportion of the surface of the sphere over which it is spread; that is, in the inverse proportion of the square of the distance. (Plate I. Fig. 4.) $S A^2 : S M^2 ::$ motion at M : motion at A. This is also the proportion in which the angular velocity of the earth about the sun varies. $S A^2 : S M^2 ::$ heat received at M : heat received at A. Therefore equal amounts of heat are received from the sun in passing over equal angles round it, in whatever part of the ellipse those angles may be situated. This is true however the ellipse may be divided by the straight line A S P. The two segments, A M P and A I P, will however be described in unequal times; but the greater proximity of the sun compensates for the more rapid description of the smaller segment, and thus an equilibrium of heat in the two hemispheres is maintained.

§ **295.** These two great causes which we have spoken of as causing the variety of seasons, are, in some parts of the world, counteracted, and in all variously modified.

In the torrid zone the vertical sun raises such vapors and causes such rains, that the season which should be the hottest, is in some places the coldest of the year. And the intermediate months, which correspond to the spring and autumn of temperate climates, are the hottest of the year. Between the tropics then we find no regular summer and winter, but rainy and dry seasons.

In polar and circumpolar regions, the days lengthen and shorten so very rapidly, that spring and autumn are unknown; vegetation advances with the utmost rapidity, and harvests ripen in the short summer, which can never be brought to maturity under warmer suns.

In the temperate zones, the change from summer to winter lasts as long as each of these seasons, and we accordingly reckon four seasons. These do not however correspond with the astronomical seasons, for causes before given.

The lengths of the astronomical seasons differ considerably, as the following table shows.

	Days.	Hours.	Minutes.
Spring lasts from the vernal equinox to the summer solstice,—	92	21	50
Summer lasts from the summer solstice to the autumnal equinox,—	93	13	44
Autumn lasts from the autumnal equinox to the winter solstice,—	89	16	44
Winter lasts from the winter solstice to the vernal equinox,—	89	1	33

The autumn and winter of the northern hemisphere are shorter than the corresponding seasons in the southern, because the perihelion is passed through in the northern winter. If the earth were in its perihelion precisely at the time of the winter solstice, the northern autumn and winter would be of equal length, and the rest of the year would be equally divided between its spring and summer. As the perihelion is 10° in advance of the winter solstice, the winter season is most shortened by the rapidity of the earth in its orbit, and the summer season includes that portion of its orbit which is performed most slowly.

§ **296.** All animals and plants have periods of repose. Some a long arctic sleep, others a slight cessation of their energies. All have periods of awakening to which their powers and habits are adapted.

While great causes bring us daily variety, a multiplicity of minor and apparently changeful causes bring us some of the most stable arrangements in nature. The climate of a place is made up of general, innumerable and local causes, which blending together, and sometimes counteracting one another, give year by year, and even for each month, almost unvarying results. Thus while we repose on the stability of this our home, we find daily something new to enjoy in its unexpected beauties.

CHAPTER XIII.

POSITION OF PLACES ON THE EARTH AND OF STARS IN THE HEAVENS.

Modes of defining position on the Earth's Surface. Methods of finding Latitude. Longitude. Its determination by the Moon's motion. The Sextant. Eclipses of Jupiter's Satellites. Determination of Local Time. Lunar Distances. The Theodolite. Celestial Globes and Maps. Apparent Motions of the Planets. The Fixed Stars. The Zodiac. The Constellations. The Milky Way. Proposed Revision of the Constellations.

§ **297.** Having ascertained the shape and dimensions of our globe, we wish to find our position on it. This may be done in two ways, by referring our position to the natural features of land and water, or by giving our latitude and longitude. Both modes of description are employed, and each has its advantages. Our latitude and longitude remain unchanged, and they furnish the shortest and most exact mode of describing our position. Latitude gives us some notion of the climate of a place. But if we knew places on earth only by their latitude and longitude, we should find it as difficult to fix them in our mind as to remember the positions of the stars. The natural features of the earth are more easily remembered, but their dimensions, and the latitude and longitude of these, must be accurately fixed before we can refer smaller places to them.

No map or chart is of much value as a representation of the earth's surface. Particular portions of it may be faithfully represented on a plane surface, but a globe gives the only correct idea of it as a whole. There are two mòdes by which a correct representation of the earth's surface may be obtained; by finding the latitude and longitude of a great number of points, and filling in the intermediate spaces by local surveys; or by finding the latitude and longitude of a few points, two perhaps in each country, and then dividing the whole country into a number of triangles. In both of these ways we must refer to the heavens for the position of our starting point.

§ **298.** The latitude of a place is easily found. It is equal to the altitude of the elevated pole. Equal differences of latitude should not however be represented by exactly equal intervals of surface, if great exactness is required. The ellipticity of the earth causes degrees of latitude to be a little longer as we approach the poles. Latitude is reckoned from the equator, and is called north or south according as the place lies north or south of the equator.

The altitude of the elevated pole above the horizon might be directly observed on the limb of the mural circle, if any bright star stood directly therein. This not being the case, a bright star near the pole, (called the polar star,) is selected, and observed in its upper and lower culminations—that is, when it passes the meridian above and below the pole. One half the sum of the star's greatest and least altitudes corrected for refraction gives the altitude of the pole, and therefore the latitude of the place.

It may be found by the observed altitude or the observed zenith distance of a star or other heavenly body when in the meridian. In observations at sea, the sun or moon is observed instead of a star, it being difficult, from the motion of the vessel, to obtain a correct observation of the meridian altitude of so small a body as a star appears. On land the inequalities of the surface make it difficult to obtain a true horizontal boundary, the zenith distance is therefore employed. The declination of the observed star being previously known, it must be added to the observed zenith distance (corrected), if both bodies are on the same side of the equator. But if the place is in north latitude, and the star has a southern declination, the declination subtracted from the zenith distance gives the latitude.

The zenith distance of a star may be obtained more accurately by making several observations on it at different altitudes, before and after culminating, when it is near the meridian. The latitude may thus be obtained within a few seconds.

If the latitude of one place is known, that of another may be found by observations on a star which passes near the zenith of both places. The calculation is more simple when both places are on the same meridian, and when both observations are made on the same day.

§ **299.** These operations are so easy in practice, and opportunities are so continually offering themselves, that the latitude of a place may generally be determined even under the most unfavorable circumstances, and its determination, by means of celestial phenomena, is the most important application of astronomy to the purposes of civil life.

But the longitude cannot be so readily found. France, Holland and England for a long time offered in vain great rewards to any one who should discover a mode of ascertaining longitude at sea. In the latter part of the seventeenth century, Flamstead gave his opinion that if we had tables of the places of the fixed stars, and of the moon's motions, the longitude might be found. Upon this Mr. Flamstead was appointed astronomer royal, and an observatory was built at Greenwich for him ; and the instructions to him and his successors were that they should apply themselves with the utmost care and diligence to rectify the tables of the motions of the heavens, and the places of the fixed stars, in order to find out the so much desired longitude at sea, for the perfecting of the art of navigation. It was not however till after Mr. Flamstead's death that the tables of the moon's motions were corrected, and an instrument invented by which altitudes could be taken at sea. The principle of this instrument is that property of reflected rays by which the angle between the first and last directions of a ray which has suffered two reflections in one plane, is equal to twice the inclination of the reflecting surfaces to one another. The instrument is called a sextant if one sixth part of a graduated circle is used, a quadrant if one fourth part. Sometimes a whole circle is used.

§ **300.** Let A B (Fig. 7, Plate I.) be the limb, or graduated arc, of a portion of a circle 60° in extent, but divided into 120 equal parts. On the radius C B let a silvered plane glass D be fixed, at right angles to the plane of the circle, and on the movable radius C E let another such silvered glass C be fixed. The glass D is permanently fixed parallel to A C, and only one half of it is silvered, the other half allowing objects to be seen through it. The glass C is wholly silvered, and its plane is parallel to the length of the movable radius C E, at the extremity E of which a vernier is placed to read off the divisions of the limb. On the radius A C is set a telescope F, through which any object Q may be seen by direct rays which pass through the unsilvered portion of the glass D, while another object, P, is seen through the same telescope by rays, which after reflection at C, have been thrown upon the silvered part of D, and are thence directed by a second reflection into the telescope. The two images so formed will both be seen in the field of view at once, and by moving the radius C E, will, (if the reflectors be truly perpendicular to the plane of the circle,) meet and pass over, without obliterating, each other. The motion, however, is arrested when they meet, and at this point the angle included between the direction C P of one object, and F Q of the other, is twice the angle A C E included between the fixed and movable radii C A, C E. The angles :—

$$PMQ + MCF = CFD,$$
$$CFD = FDN = GCD + GCF,$$
$$PMQ + MCF = GCD + GCF;$$
$$GCD = RCP,$$
$$RCP = MCF + GCF = GCD;$$
$$PMQ + MCF = MCF + 2\,GCF,$$
$$PMQ = 2\,FCG = 2\,ACE.$$

Now the graduations of the limb being purposely made only half as distant as would correspond to degrees, the arc B E, when read off as if the graduations were whole degrees, will, in fact, read double its real amount, and therefore the numbers to read off will express not the angle A C E, but its double, the angle subtended by the objects.

As the sextant can be held in the hand and requires no fixed support, it is of great use in nautical astronomy. It not only measures the distance between two stars, or the moon and a star, but gives the altitude of the moon or stars. For altitudes at sea, as no level, plumb-line, or artificial horizon can be used, the sea offing affords the only resource. The image of the sun observed, as seen by reflection, is brought to coincide with the boundary of the sea as seen by direct rays. Thus the altitude above the sea-line is found; and this corrected for the dip of the horizon gives the true altitude of the sun. On land an artificial horizon may be used, and the consideration of the dip is rendered unnecessary.

§ **301.** Longitude is reckoned from a point arbitrarily chosen, and is called east or west according as it is within 180° east or west of the meridian which passes through this point. It is sometimes reckoned westward all round the globe, to correspond with the reckoning of right ascension. Celestial longitude may be expressed merely in degrees, or in signs, or the sign in which the star lies may be mentioned. Thus longitude 45° is either expressed thus, or as the 15th degree of Taurus, which is written thus,—♉ 15°, or as 1s., 15°; the 1s., or one sign, being taken merely as a mode of expressing 30° longitude.

The longitude of a place is found by means of its local time. Every place has its own sunrise, its own noon. All places under the same meridian are brought under the sun and have their noons at the same time. All places not under the same meridian have their noons at different times. The astronomer regulates his clock to indicate 0h. 0m. 0s. when the vernal equinox comes on to his meridian. He must therefore mention not only at what hour an event happened, but at what hour of what local time it happened.

Suppose two observers to set and regulate their chronometers each by his own true sidereal time. If one of these chronometers were transported and compared with the other, they would differ by the time occupied by the equinox, or by any star, in passing from the station of one observer to that of another; in other words, by their difference in longitude expressed in sidereal hours, minutes, and seconds.

If chronometers were perfect no better mode of ascertaining longitudes than this need be desired. An observer provided with such an instrument might, by journeying from place to place, ascertain the differences of longitude with great precision.

§ **302.** If he travels westward his chronometer will appear to gain, though it really goes correctly. Suppose he sets out from A, when the equinox is on the meridian, or his chronometer at 0h., and in 24 hours sidereal time has travelled 15° westward to B. At the moment of his arrival there his chronometer will again point to 0h.; but the equinox will be, not on his present meridian, but on that of A, and he must wait one hour more for its arrival on that of B. When it does arrive there, his watch will point not to 0h., but to 1h., and will therefore be 1h. fast on the local time of B.

If an observer travels westward, and adopts the local time of each place he reaches, he loses an hour for every 15° he advances. His watch, which shows the number of hours which have actually passed, is an hour fast. He throws away that hour, and considers that he has travelled one hour less than the true number. If he travels entirely round the globe, he will have suppressed an hour twenty-four times, and thus will reckon one day less than if he had remained stationary. Each of his days will have been a little longer than to a person at rest, and he will actually have seen the sun rise once less often than if he had remained at home. We have 365 days in a year, because the earth carries us round under the sun 365 times a year. By travelling westward, one of these turns is gradually, for that observer, cancelled, and he has but 364 days. Travelling eastward, a chronometer is found slow; for every 15° of advance the hours are called one more than the true number,—one turn is added to the earth's 365 turns, one day is gained in a year. This has actually happened to navigators. Two settlements on the same meridian may in this way differ a day in their reckoning of time according as they have been colonized by settlers arriving from the eastward or from the westward.

§ **303.** Instead of comparing two local times by a chronometer, the instant of the occurrence of a *phenomenon* may be noted in the local time of two places differing in longitude, and their difference may be thus ascertained. This *phenomenon* may be a natural or an artificial signal. Natural signals, such as eclipses of the moon, may be seen at the same instant of time over a hemisphere; artificial signals are visible to a much less distance. The exact local times of two stations being known, a signal of some definite kind, as a flash of powder, is made between the two within sight of both. Since light is so very swift, the signal will be seen at the same absolute instant at both places, and the difference of their local times gives the difference of their longitudes. A line of such signals is sometimes used. The distance to which they may be seen is very considerable. Over the sea the explosion of rockets may be seen fifty or sixty miles; and in mountainous countries at much greater distances. Meteors, which are natural signals, may also be used to determine longitude. From their great height they are seen over a large extent of country, and their sudden appearance makes it easy to seize the moment of appearance. The magnetic telegraph has also been used for finding longitude. Signals are transmitted, apparently instantaneously, from one station to another, and the difference of the local times of the two stations gives the difference in longitude.

Another natural signal is an eclipse of Jupiter's satellites. These eclipses have a great advantage; the time of their occurrence at any fixed station, as at Greenwich, can be predicted with such accuracy as to stand in the place of a second observation. An observer may compare his local time with the predicted Greenwich time, and thus learn his longitude. This mode is not however susceptible of great exactness, since the moment of the commencement of the eclipse cannot be seized; neither can it be employed in navigation, because the rolling of the ship prevents nice telescopic observations.

Lunar eclipses have likewise been used for obtaining longitude, but they are liable to the same objection to a still greater extent. An error of a minute in this observation would cause an error of a quarter of a degree of longitude, and in fact a much greater error of observation is unavoidable.

§ **304.** The most simple and exact method of finding the time at a given place, is to observe the instant when the limb of the sun, or a star of known right ascension, is on the middle wire of a transit instrument properly adjusted. At that moment the star is on the meridian; its right ascension expressed in hours, minutes and seconds is the sidereal time. No method is equal in accuracy to this method of transit; but as it can scarcely be employed except in fixed observations, it is necessary to adopt some other more generally applicable. By means of a sextant the altitude of a known star may be taken, the time being carefully observed, while the star is at a considerable distance from the meridian; when the star has passed the meridian, and is at the same altitude on the other side, let the time be noted carefully. Since the apparent altitudes are equal, refraction is the same, and the true altitudes are equal. The instant of the star's being on the meridian will exactly bisect the interval of the observations.

§ **305.** The method of lunar distances is more useful than any other in finding the longitude. Its principle is identical with that of lunar eclipses and eclipses of Jupiter's satellites. The object is to find some celestial phenomenon which may be observed under different meridians, and by which the two observers may compare the times they reckon at the same absolute instant. The hour at which the phenomenon will happen may be calculated for one meridian, and observed for the other, and the comparison of the observed with the calculated time is substituted for that of two observed times. The face of the heavens has been compared to the face of a dial, on which well known bright stars are the marks which give the time, the moon is the hand, and Greenwich time is 12 o'clock, or the starting point. The marks on the dial are not at equal intervals, but the intervals are known; the moon varies in rapidity of motion, but her variations are known ; the observer does not stand in the same plane with the centre of the moon and the centre of the earth, but he can allow for the change this causes in the moon's apparent place. The moon is selected for the index rather than the sun or any planet, because her greater rapidity of motion leaves less uncertainty about the precise moment corresponding to any given angular distance from the stars. The sun's apparent motion is only one thirteenth of that of the moon; and if his place were determined with an uncertainty of only a quarter of a minute in space, this would leave an uncertainty of nearly six minutes of time in the longitude.

§ **306.** It is evident that there is only one particular instant at which the moon can be at a certain distance from any fixed star. If this distance is ascertained at any moment, and the Greenwich time at the moment she had the same difference is counted from the tables, the longitude is found. The true distance of the moon's centre from the star, when corrected for refraction and parallax, is the same for every meridian. The British Lords Commissioners of the Admiralty publish annually a Nautical Ephemeris, containing the distance between the moon and certain bright fixed stars near her path, for every three hours.

We have thus, in lunar distances and in chronometers, two independent methods of finding the longitude, each of which may act as a check on accidental errors in the other. Several chronometers may likewise be used to correct one another. The rate at which a chronometer gains or loses should always be verified before leaving some known meridian. And if a vessel remains long enough in port, its rate should again be ascertained, since change of temperature or other causes may have altered it. If the rate is accurately known, and the error allowed for, chronometers will give the longitude throughout a long voyage.

§ **307.** In determining the relative position and distances of places on land, use is made of an instrument called a Theodolite, which resembles the altitude and azimuth instrument; its use has been thus described. It is evident that, as every object to which the telescope of a theodolite is pointed has some certain *elevation*, not only above the *soil*, but above the level of the *sea*, and as, moreover, these elevations differ in every instance, a *reduction to the horizon* of all the measured angles would appear to be required. But, in fact, by the construction of the theodolite, this reduction is *made* in the very act of reading off the horizontal angles. Let E (Fig. 5, Plate I.) be the centre of the earth; A, B, C, the places on its *spherical surface*, to which three stations, A, P, Q, in a country, are referred by radii E A, E B P, E C Q. If a theodolite be stationed at A, the axis of its horizontal circle will point to E, when truly adjusted, and its plane will be a tangent to the sphere at A, intersecting the radii E B P, E C Q, at M and N, *above* the spherical surface. The telescope of the theodolite, it is true, is pointed in succession to P and Q; but the readings off of its azimuth circle give, *not* the angle P A Q, between the directions of the telescope, or between the objects P, Q, as seen from A; but the *azimuthal angle* M A N, which is the measure of the angle A of the spherical triangle B A C. The sum of the three observed angles of any of the great triangles in geodetic operations is always found to be rather more than 180°; and this *excess*, which is called the *spherical excess*, is so far from being a proof of incorrectness in the work, that it is essential to its accuracy, and offers at the same time another palpable proof of the earth's sphericity.

§ **308.** The true way, then, of conceiving the subject of a trigonometrical survey, when the spherical form of the earth is taken into consideration, is to regard the net-work of triangles with which the country is covered, as the bases of an assemblage of pyramids converging to the centre of the earth. The theodolite gives us *the true measures of the angles included by the planes of these pyramids;* and the surface of an imaginary sphere on the level of the sea intersects them in an assemblage of spherical triangles, above whose angles, in the radii prolonged, the real stations of observation are raised, by the superficial inequalities of mountain and valley. These triangles may afterward be reduced to the level of the sea, by applying the rule for the spherical excess, and the ellipticity of the earth may also be taken into account in very nice surveys.

The irregularities of the earth's surface are learned by sounding the sea and by applying the barometer to the air, or by direct measurement of heights. The pressure on the barometer at any height informs us in what stratum of the air that height is, and consequently how much it is above the sea level.

§ **309.** Celestial globes are more easily constructed than representations of the earth. A concave surface would of course be the most perfect representation of the heavens, but as we cannot have one so large that the spectator can stand within it, convex globes are generally used. Triangles may be transferred from the heavens to this surface, bright stars taking the place of stations in terrestrial triangles. A better way to construct a globe is, as the earth rotates to observe the place in the heavens of each celestial object which passes our meridian, and to refer it to its place on an imaginary sphere conceived to revolve with the stars. By observing both in a north and south latitude the whole sphere may be mapped out, and their true places assigned to the fixed stars. As on the earth's surface we may refer points to the natural portions of the earth, or to latitude and longitude, so in the heavens we may describe a star by its situation relative to others, by its belonging to some constellation, or by its right ascension and declination. The features of the earth probably change in long ages, continents rise and fall, mountain chains are thrown up, gulfs open. We have learned that apparently slight changes constantly occurring may, in equally long or perhaps longer intervals, alter the appearance of the heavens.

Terrestrial latitude and longitude we have every reason to suppose immutable. But this is not the case with declination and right ascension. As on the earth the meridian of Greenwich or of some other place is selected as the point from which longitude is counted; so in the heavens, the vernal equinox, one of the points in which the equator and ecliptic intersect, is chosen for the zero point of right ascensions. If the axis of the earth is, as we have called it, absolutely immovable, this point of intersection will always remain the same, and right ascensions and declinations, celestial latitudes and longitudes, will remain unchanged. If by any cause motion of the earth's axis is induced, the equinoctial must share this motion, and must intersect the ecliptic in a different place. Or if the position of the ecliptic among the stars should, by any cause, be changed, the points of intersection would in like manner be moved. This leads us to the inquiry, which will be answered in the next chapter, whether there are in the heavens any perturbations.

§ **310.** Maps of the heavens constructed by an inhabitant of the earth, will be in their natural features correct for an inhabitant of any of the planets, but not in their arbitrary references. The constellations and the milky-way are the same to all dwellers in our system, but each planet would have its own ecliptic and equator, its hour-circles, and lines of longitude to which to refer stars.

We have assigned no place on the celestial globe to comets, to the planets, or to the sun and moon, which often, from their apparent wandering, are included among the planets. Since the planets change place continually, their position is better described by giving their distance from the sun. As the sun is the centre of our system he could not very well appear on a map or globe, and as we transfer our motion to him he also occupies different positions in the heavens. Perhaps it will be well to complete our map of the heavens by mentioning to what portions of it the planets are confined, and how their motions in that portion appear.

§ **311.** Owing to the earth's motions, the sun, as seen from the earth, covers in the course of the year the celestial zone, extending from 23° 28′ north declination, to 23° 28′ south declination. It covers every portion of this zone in the course of the year, its path one day partially lapping over its path the preceding day. It never departs from this zone north or south.

The moon really moves round the earth in a kind of spiral, so that her disc at different times passes over every point in a zone of the heavens, extending rather more than 5° 9′ on each side of the ecliptic. At one time or other she occults every planet and star within this space. The occultation of a star by the moon is not more frequent than by the sun, but the dimness of the moon allows it to be seen. The moon seems to pass over the star, which instantaneously vanishes at one side of her disc and after a short time as suddenly reappears on the other.

The sun and the moon are so important to the earth that the daily apparent course which the earth's rotation impresses on them is noticed. This is from east to west, while the moon's actual monthly and the sun's apparent yearly course are from west to east. The earth likewise impresses daily motions from east to west on the planets, but these are left out of consideration in tracing the courses of the planets.

No eye but one placed entirely outside of the system, or at the sun, could see the motions of the planets truly. Since their paths are nearly in the plane of the ecliptic, we see their motions not in plane but in section ; their real angular movements and linear distances being all foreshortened and confounded undistinguishably, while only their deviations from the ecliptic appear of their natural magnitude, undiminished by the effect of perspective.

To an observer whose point of view is itself in motion, the paths of the planets are transformed into zigzag lines ; they appear now to advance rapidly, now to stand still, and then to recede. The planets nearer to the sun than the earth are called inferior, those more distant than the earth are called superior planets. There are some differences in the motions of the superior and the inferior planets as viewed from the earth.

§ **312.** The inferior planets appear to vibrate each side of the sun, never removing far from him, and advancing with him in the ecliptic. An inferior planet has two conjunctions, a superior one when it is beyond the sun with regard to the earth, and an inferior one when it is between the sun and the earth. Superior planets have one opposition and one conjunction. When an inferior planet passes beyond the sun, with regard to the earth, its motion, compared to the stars, appears direct. When it passes between the earth and the sun, its motion appears retrograde. In transition from one of these states to the other, its motion is imperceptible, and it appears stationary. These appearances would take place if the earth were stationary. The earth's motion only modifies them by changing the points at which they become stationary, and by making them appear stationary longer. Conjunction also takes place less frequently in consequence of the earth's moving in the same direction with the planets. While the planet has performed half of its orbit, the earth has advanced, and the planet must also advance to be in conjunction with the sun. Thus conjunction of the inferior planets takes place rather less than twice in one of their years. The planet appears to vibrate a few degrees on each side of the sun, but never appears in any remote part of the heavens.

§ **313.** The superior planets appear more capricious in their movements. Let us suppose ourselves on a great lake in a boat, while another boat passes in the same direction between us and the shore. If both move with equal quickness, the boat nearest the shore will appear motionless. If the shore is at a great distance, the external boat will answer for a long time to the same objects. If one of the superior planets represents the boat nearest the shore, and the sphere of the fixed stars an infinitely distant bank, while the earth and this planet move with equal quickness and in parallel directions, the latter will appear to us to answer to the same point of the heavens, to be stationary. If the boat nearest the shore goes less swiftly, the spectator in the other will see it retrograde, and hide successively along the shore objects in the direction opposite to that in which he goes. When, on the contrary, the former goes more quickly, it will appear to advance in the true route and directly. If the one farther from shore moves in an opposite direction, the nearer motion will still be direct and very rapid. All this represents what happens to the superior planets with regard to the earth. When the latter surpasses them in quickness in the same direction, they are retrograde; they become direct when the earth is left behind or is in that part of her orbit which has a contrary direction. Thus the stations and retrogradations of the superior planets are toward oppositions ; they are direct in the rest of their course.

§ **314.** When a superior planet approaches its opposition there must be a little arc where two lines joining the ends of the planet's arc with the ends of the earth's are parallel. During this time the planet is stationary, for being referred to the immensely distant fixed stars, the distance between the two ends of the arc is not perceived, and the two positions are confounded. Afterward the earth surpasses the planet in quickness, its motion becomes retrograde, and opposition is made toward the middle of the arc of retrogradation. At last the retrogradation is stopped by the planet's becoming stationary. It then becomes direct, and remains so as long as the earth passes through that part of her orbit where she moves in an opposite direction from the planet. The motion of the planet accelerates continually from its last stopping-place until it is hidden in the rays of the sun, and from the moment when it disengages itself, it slackens its swiftness till the next station.

The motion of the earth alters the motions of the superior planets far more than those of the inferior. If it were at rest the planet would mark out on the sphere a circle, and have one opposition and one conjunction in each of its own years. As it now is, Jupiter, in each of his revolutions, is twelve times overtaken and passed by the earth, Saturn thirty times, and they have respectively twelve and thirty oppositions and conjunctions. Their paths among the stars are zigzags, retrograding as many times as the planet has oppositions in a year. Thus their apparent often far outstrips their real motion. Herschel does not move through 5° of his orbit in a year, yet to us he often appears to move much more rapidly. On the whole, however, the amount of direct motion more than compensates for the retrograde ; and by its excess the gradual advance of the planet from west to east is maintained. Saturn appears longer in turning than Jupiter ; Jupiter than Mars ; Mars than Venus. For a week together the most powerful telescope scarcely shows any change in the situation of Saturn; during a whole month he moves less than a degree.

It rarely happens that more than one or two planets are in the same part of the heavens at the same time. More than 2,500 years before our era, the five great planets were in conjunction.

§ **315.** The appearances of the planets convinced Copernicus that the earth was not, as then believed, the centre of the system. He observed that Mars and Jupiter and Saturn appeared much larger in their oppositions than in their conjunctions. Thus they either had not the earth for a centre, or had prodigious excentricities. If the sun were the centre, the change of size would be simply accounted for. He would not publish his discovery till he had satisfied himself that it accounted for all the details of their motions. After thirty-six years, with great reluctance, he gave it to the world, not as a physical truth, but as a convenient hypothesis.

Not only the variations in size, but all apparent irregularity of the planets' motions vanishes as soon as they are referred to the sun as a centre. Their periodic times, their distances and velocities, in short all the elements of their orbits may be found. All these must be found to give the position of a planet in space; finding its latitude and longitude only give its place (among the constellations) on the celestial vault.

§ **316.** The latitudes and longitudes mentioned thus far have been reckoned from the earth as a centre, and for the fixed stars they are the same as if reckoned from the sun as a centre. But the places of the sun, moon and planets referred to a sphere having the earth for its centre differ from their places on a sphere concentric with the sun. Their longitudes and latitudes found on the former sphere are called geocentric, those on the latter sphere are called heliocentric. If the sun is the centre, the longitudes are calculated from the vernal equinox. The sun's and the earth's latitude are always nothing. The heliocentric longitude of the earth equals the sun's geocentric longitude$+180°$, or vice versa. The heliocentric equinoxes and solstices are therefore the same with the geocentric. The geocentric place of the moon or of one of the planets being known, and the earth's distance from the sun and from the moon or planet being known, the moon's or planet's heliocentric place may be found. Or if you wish to calculate the geocentric place of a planet, let its heliocentric place and that of the earth be known; the angle at the sun is the difference between these two, and completing the triangle we have the place of the planet with regard to the earth. Sometimes the heliocentric longitude precedes the geocentric, sometimes it follows it, and sometimes coincides with it..

If the earth moved in a circle with a uniform velocity about the sun placed in the centre, its position at any time with regard to the line of equinoxes could easily be calculated; for as one year is to the time elapsed, so would 360° be to the arc passed over. The longitude so calculated is called the mean longitude. But since the earth's orbit is neither circular nor uniformly described, this rule will not give us its true place in the orbit at any particular moment. Still the true place differs very little from the place so determined, called the mean place, and may always be found from it by applying a correction, or equation as it is termed, whose amount is not very great, and depends on the equable description of areas about the sun. The proportion is; as one year : the time elapsed : : the whole area of the ellipse : the area of the section swept over by the radius vector in that time.

§ **317.** The quantity by which the true longitude of the earth differs from the mean longitude is called the equation of the centre, and is additive from perihelion to to aphelion, beginning at 0°, increasing to a maximum, and again diminishing to zero at the aphelion. After this it becomes subtractive, attains a maximum, and again diminishes to 0° at the perihelion. Its maximum, both additive and subtractive, is 1° 55′ 33″.

The maximum value of the equation of the centre depends only on the ellipticity of the orbit. If then the former inequality can be ascertained from observation, the latter may be found from it. The sun's exact longitude may be ascertained for every day, and compared with the mean longitude, and the greatest amount of its defect or excess ascertained. This is a more accurate mode of learning the excentricity of the orbit than that of concluding its distance from its apparent diameter. Since the true and mean longitudes agree twice a year, but do not agree from day to day, the true must, during a part of each half year, increase more rapidly, and during a part less rapidly than the mean. The earth, starting from perihelion, describes each day arcs which exceed $\frac{1}{365}$ part of its orbit; as it approaches the aphelion, the arcs described fall short of $\frac{1}{365}$ of its orbit. But the true longitude each day measures the earth's whole advance from the perihelion. If it gains a great deal the first day, and less the next, and less the one after, and so on, the whole amount added to the mean longitude will for some time increase, but a time must come when it will diminish; since the orbit is nearly circular, this point is near quadrature. We shall find hereafter that the longitude of the earth's perihelion has a very slow advance on the ecliptic. This of course causes the equation of the centre to be additive in different portions of the ecliptic in different centuries.

§ **318.** From whatever spot on earth we look into the heavens, we can never, even with telescopes of the highest power, discover any limit; nothing intervenes to check our sight. The heavens assume the form of a hollow sphere, of which our eye is the centre, because the eye seeing equally far in all directions, refers all it beholds to a concave surface every where equally distant from itself. A hollow sphere is the only surface which answers this condition, and the eye maps the stars down on this sphere with reference merely to their direction from us; the faintest may be millions of miles beyond the brightest; their actual distance from us does not affect their place on the sphere. In the same way, when we are painting a group of persons, each one takes his place on the background formed by the wall without regard to his distance from us; and just so we refer all the buildings, trees and figures in a landscape to the mountain which closes in our horizon.

But when we look into space, there is no mountain, no wall, to shut in our horizon; our view is only limited by the extent of our power of vision. This power differs extremely in different persons, and no one is conscious where it ends, though all are conscious of the apparent shape which arises from its ending. The radius of our sphere of vision includes the milky light of stars in other clusters. All things within this distance, whether thousands, millions, or millions of millions of miles distant, are referred to one same imaginary surface. Only one half of this sphere is visible to one observer—the earth on which he stands excluding from view the hemisphere beneath his feet. By means of observations in different latitudes, and at all seasons of the year, the whole surface of this concave sphere is however familiar to us, and is recognized by means of its natural features, the stars.

§ **319.** But though our only landmarks, the stars are at first view like the sands of the desert or a snowy plain. We are however not wholly without signs. We notice striking differences in the size, color and arrangement of the stars among themselves; we find their arrangement never varies; that stars which have to our fancy taken the form of a bear, retain it whether the bear have his feet or his head toward our horizon. We begin to map out the heavens, and since we find nothing like those outlines of land and water which define the surface of our globe, we invent, or rather some Chaldean shepherds invented for us, some thousands of years ago, odd, uncouth figures which bear a slight or fancied resemblance to the forms of the groups of stars. Wherever a few bright stars lay sufficiently near together to assume the form of a Lyre, a Swan, a Virgin, or any other figure, they were called a constellation, and took their name from their fancied form. Thus in time the whole celestial vault has been covered with imaginary figures which serve well enough to indicate to the common inquirer in what part of the heavens a star is placed, and which have usually some mythological or historical interest which aids the memory. Probably the most striking stars and groups were the first to receive a name. The natural event most important in its effects, and most obvious to an uncultivated people, is the return of the seasons. This must have been early perceived to be accompanied by an apparent motion of the sun among the stars. If the light of the stars were much stronger, or that of the sun much weaker, we might see him pass by the stars in each part of the ecliptic, as we do the moon. But his path was easily ascertained by observing what stars rose or set with him each night, or what were opposite to him at midnight.

§ **320.** It must have been early observed that the planets and the moon never wandered far from the apparent path of the sun. All the motions of the planets then discovered, are performed within a zone extending about 8° each side of the earth's path. The stars in this zone were very early formed into constellations, and considered as resembling figures of animals. The name Zodiac, from a Greek word signifying animal, was consequently applied to this zone. The constellations which lie in it are by us called the zodiacal constellations. It has been supposed that the country and period in which they were named might be ascertained by calculating in what country the agricultural operations of which the signs are symbols would coincide with the presence of the sun in that constellation. Some antiquarians have inferred that Egypt is the country; but as the Egyptians borrowed their mythology and perhaps their civilization from some Oriental people, it seems more probable they received their astronomical calendar from the same source. And as the Hindoo and other Oriental nations show some traces of a similar division, it is probable that they and the Egyptians received them from the same more ancient source. Perhaps the Egyptians received the division as a loose one, and first made it definite; for several coincidences make it probable that it received its present arrangement in Egypt.

Owing to causes hereafter to be explained, the sun does not appear among the same stars at the same season in which he did centuries ago. If we consider the zodiacal constellations as symbols connected with husbandry, the agreement could only have subsisted when the sun was in the constellation Aries on the 21st of March.

§ **321.** We may then suppose the constellations Aries,

Taurus and Gemini, to have been named from the young of animals being added to the flock in spring. After this the sun seems to retreat toward the south, and the next constellation is called Cancer, from the crab, which moves backward. Leo indicates the violent heats of summer, and Virgo represents a gleaner, and the time of her appearance coincides with harvest time in Egypt. The perfect equality of the days and nights in the next month is symbolized by Libra, the balance. The diseases produced by the departure of the sun gave to the next sign the name of Scorpion, because it is mischievous, and was thought to have a sting in its tail. The next month was the season for hunting, and had for its emblem Sagittarius, an Archer. In the next month the sun appeared to ascend from the south toward the equator, and it had for its emblem Capricornus, because the goat is accustomed to ascend the highest points of ground. The next sign is Aquarius or the Watercarrier, named from the rains that generally fall at this season, or from the inundation of the Nile. And the last sign is Pisces, the Fishes, so called perhaps because they were thought to be most fit for use at that time.

If the names were given to the constellations in which the sun then was, the antiquity of 15,000 years is required for the zodiac. Perhaps they were given not to those in which the sun then was, but to those which were opposite to him, and which consequently were rising at sunset at any given spot. This theory brings down the invention of the constellations to about 2,500 years before Christ; it has been adopted by La Place and several distinguished philosophers.

The Greeks probably received their astronomical knowledge from the Egyptians. It is evident they did not themselves name the constellations, because they could not, for some time at least, explain them according to their own mythology. Probably out of the medley of men, animals, and other objects, with which earlier astronomers had filled the heavens, they selected and retained the figures which suited the deeds of their own heroes and deities. Thus Aries is supposed to represent the ram, whose golden fleece was the object of the Argonautic expedition; Taurus, the bull which was tamed by Jason; Gemini, Castor and Pollux. The ship among the southern constellations is supposed to be the Argo; and the Ursa Major, which to the Greeks would never set, is the nymph Calisto, whom Juno forbad Oceanus to receive into his bosom. When the Scorpion appears in the east, Orion must sink beneath the western horizon, because Artemis, to punish the audacity of the mighty hunter, sent a scorpion which bit him in the heel.

§ **322.** When it was found convenient to divide the ecliptic into twelve equal parts of 30° each, it was found that each of these parts would be in the neighborhood of one of the zodiacal constellations. For these are twelve in number, and situated at nearly equal distances one from another. The ecliptic was thus divided; and each portion was named for the constellation which was near it; and the divisions themselves, and also the constellations which gave name to them, were called the signs of the zodiac, and characters were invented to express them. The names of the constellations, or signs, and the characters used to express them, are as follows:—Aries, or the Ram, ♈; Taurus, or the Bull, ♉; Gemini, or the Twins, ♊; Cancer, or the Crab, ♋; Leo, or the Lion, ♌; Virgo, or the Virgin, ♍; Libra, or the Balance, ♎; Scorpio, or the Scorpion, ♏; Sagittarius, or the Archer, ♐; Capricornus, or the Wild Goat, ♑; Aquarius, or the Water-carrier, ♒; and Pisces, or the Fishes, ♓.

When these constellations were fixed on to determine the names of the subdivisions of the ecliptic, the vernal equinox was a point very near the Ram, the summer solstice was near the Crab, the autumnal equinox was near the Balance, and the winter solstice near the Wild Goat. The vernal equinox therefore received the name of the first point in Aries, which it still retains; the autumnal equinox was the first point in Libra; and the tropics were called the tropics of Cancer and Capricorn. These names still continue in use, though the circumstances from which they took their origin have ceased to exist. The vernal equinox now takes place in the constellation Pisces, the autumnal equinox in Virgo, the summer solstice in Gemini, and the winter solstice in Sagittarius; and the other constellations from which the signs are named, have also changed their situation on the circle of the ecliptic. Not only, however, do the points of the equinoxes and the solstices retain their names, but the whole ecliptic is still divided into twelve portions, which are called signs, and retain the names of the constellations for which they were originally called. These signs, or portions of the ecliptic, continue to be measured at intervals of 30° each, from the actual position of the vernal equinox; the equinox having retreated, the signs or constellations of the zodiac no longer correspond with them. The constellation of the Ram is now near the sign ♉ of the ecliptic; that of the Lion near ♍; that of the Waterman near ♓. Hence we must carefully distinguish between the signs of the zodiac, which are fixed with respect to the equinoxes, and the constellations, which are movable with respect to these points.

The ancient Greeks reckoned only forty-six or forty-seven constellations. Hipparchus added Equuleus. The Hair of Berenice and Antinous afterward made the number fifty.

§ **323.** In the fifteenth century, when navigation was extended beyond the equator, and sailors noted those stars in the southern hemisphere which were not visible to the ancients, they found it convenient to group them into constellations. They did not however adapt them to the Greek mythology, but selected principally such objects as presented themselves in the newly discovered countries. Whence we have for the southern constellations, the Phoenix, the Toucan, the Little Water Snake, the Sword-fish, the Flying-fish, the Fly, the Chameleon, the Bird of Paradise, the Peacock, the Indian, and the Crane.

The ancients took only those parts of the heavens as the ground-work of the constellations where the bright stars existed. Consequently in many places there were no constellations, and the stars which were scattered over such places were called *informes*. Some of these empty spaces were very great, and contained here and there stars which were as much entitled to be formed into constellations as those of several existing ones. Therefore modern astronomers named the new constellations, called the Camelopard, the Unicorn, the Fly, and the rivers Jordan, Euphrates, and Tigris. In the latter part of the seventeenth century, the rivers were rejected, and instead of them, and in some other vacant spots, were introduced the Hounds, Mount Menalus, Cerberus, the Fox and Goose, the Lizard, the Shield of Sobieski, the Lynx, the Little Lion, the Little Triangle, and the Sextant, and also the Bow and Arrow of Antinous.

Many other names of constellations were added as compliments to monarchs or patrons, or as commemorative of interesting events or distinguished men. These often replaced former constellations, and speedily disappeared from the maps, and many are not known at the present day.

§ **324.** A natural feature in the heavens, more marked than any of the constellations, is the milky-way. This has not improbably presented the same appearance and kept the same position ever since the creation of our cluster. It traverses the constellations Cassiopeia, Perseus, Auriga, Orion, Gemini, Canis Major, and Argo, where it appears most brilliant. It then passes through the feet of the Centaur, the Cross, the Southern Triangle, and returns towards the north by the Altar, and the tail of the Scorpion, where it divides into two branches. One branch passes through the tail of Scorpio, the bow of Sagittarius, Aquila, Antinous, Sagitta, and the Swan. The other branch passes through the upper part of the tail of Scorpio, the side of Serpentarius, Taurus, Poniatowski, the Goose, and the neck of the Swan, where it again unites with the other branch, and passes on to the head of Cepheus. Here the branches unite, after remaining separate for the space of more than 100°. There is another small separation of the milky-way between Cassiopeia and Cygnus. In some parts this zone is ten or fifteen degrees broad, as in the southern parts of Scorpio, Ara, and the Cross; in others, as between Perseus and Auriga, it is not more than five degrees in width. Some parts of it are visible at all seasons of the year. In northern latitudes it is most conspicuous from July to November. It is most brilliant in the southern hemisphere.

Instead of a confused milky light, it is there more studded with brilliant stars.

§ **325.** One would suppose that nearly eighty constellations were quite enough for all useful purposes. But in the eighteenth century twenty-six more were added to the number. This extravagant number of new constellations, some of which were formed of scarcely visible stars, by no means made the study of astronomy more easy, but on the contrary confused it, and rendered it difficult. Moreover the new constellations are unsuited to the others, and chosen without taste. Astronomical instruments have some claim to a place in the heavens; but figures like the Chemical Furnace, the Easel, the Air-pump, the Printing Press, and the Electrical machine have no natural relation to the sky.

It is desirable that the heavens should be freed from so tasteless and useless an accumulation. In doing this, uniformity must be secured for the maps. The same constellations must be retained in all, and the same stars should be placed in the same parts of the figures. The forms chosen should be beautiful and pleasing, their outlines should be definite, and when once adopted should remain unchanged. It would be well to avoid similarity of names in the constellations. We have now an Ursa Major and Minor, three Triangles, Pisces and Piscis, Telescopium repeated three times, &c.

§ **326.** According to the present system, some constellations are so extensive that they exhaust three or more alphabets, and therefore it is necessary, beside the letter of the star, to give its right ascension and declination.

The largest stars of each constellation are named by the letters of the Greek alphabet, beginning with the brightest and proceeding in order. The stars next in brightness are numbered according to the Roman alphabet, and sometimes a third alphabet of Italian letters, or one numbered is required, as a^2; or numbers alone are used.

The letters do not indicate the magnitude of the stars they represent, but merely the relative magnitude of those in the same constellation. Thus α Virginis is a star of the first magnitude; α Libræ, a star of the second magnitude; and η Aquarii, a star of the third magnitude.

Among the most conspicuous constellations in the northern hemisphere are the Lesser Bear, in the direction of which the north pole of our earth continually points, the Great Bear, which is more distant from the pole, Perseus, Cassiopeia, Lyra, Hercules, the Wagoner, Orion chasing the Pleiades and Hyades, while the Dog Star, though in the southern hemisphere, follows in the train. The southern constellations are more brilliant. The Southern Cross, the Argo, the Southern Triangle, the Centaur, and the Southern Crown, are among the most splendid.

A new system of arrangement and nomenclature has been proposed, in which the heavens should be covered by a net-work of imaginary circles crossing each other at regular intervals, so that each star could be referred to its exact place. Meanwhile much confusion must exist where 3,487 stars are to be formed into 94 figures whose outlines are imaginary and undefined, neither coincide with the position of the stars nor are definite in themselves, and seem made uncouth and perplexed purposely to baffle the observer.

The zodiacal constellations which we have described are 12 in number, and contain 1,016 stars. The northern constellations are 34 in number, and contain 1,444 stars. The southern constellations are 47 in number, and contain 1,027 stars. A large number of these are telescopic stars, but they are well known.

These constellations should be seen on a map or a concave celestial globe. Such a globe represents them as they really are, and a convex one reverses their appearance to us.

§ **326.** In a celestial map the eastern part of the heavens is toward the left hand, the western part toward the right. If we stand on the earth with our face toward the north, we have the eastern part of the earth and of the heavens on our right hand. But if we face these heavens, and make a map of them, the eastern part must be depicted on the left hand. The eastern part of the heavens is that you reach by travelling eastward, and when you face it it appears on the left hand.

In all drawings of the celestial motions in this book, the west is to the right hand, the east to the left hand. On maps of the earth, the east and west points are differently placed. In looking at the heavens we look at a concave sphere, in looking at the earth we have a convex sphere, and hence the maps which represent portions of each are differently made. For the concave sphere the west is on the right, and the east on the left hand. For the convex sphere the east is on the right, the west on the left hand.

We must remember that north and south are points, the imaginary ends of the pole of rotation, while east and west are directions. This difference exists both in the earth and the heavens. One definite part of the globe is called the northern point, but no point or part is called the eastern in this sense. If we travel north on the surface of the globe we come to this point, where we can no longer go north, but must turn and go south. But we can travel east till we return to our starting place, nay we might go round the globe again and again and yet always travel eastward. Thus in the heavens the earth moves always in an east direction. There is no part of the heavens which is called the eastern, none which is called the western part.

CHAPTER XIV.

LAWS OF SHAPE AND MOTION.

Attraction of Gravitation. Effect of Gravitation on the figures of the Sun and Planets. The Figure of the Earth that of Equilibrium. Illustration of the effect of Rotation on a Fluid Mass. Laws of Gravity. Centre of Gravity.

§ **328.** In the preceding pages many individual facts have been stated. The form, relative masses, and orbits of the members of the solar system have been given. We would now study the laws which decide these forms, and govern these motions. Before doing this we must make ourselves familiar with a principle whose workings pervade all we yet know of the universe, a principle which influences the form and motions of all matter devoid of life,—the principle of gravity. Of the nature of gravity we know nothing; we call it an attraction because in obedience to it bodies approach one another. But though we are ignorant of its nature and of its mode of action, we know with certainty the results to explain which we suppose its existence, and we know that they take place invariably.

Every particle of matter, as far as we know, attracts every other particle. If the particles are in a fluid or gaseous state, and are hindered by no other force, they rush together, and take the form which satisfies their mutual attraction. In this way we account for the form of the celestial bodies. If the bodies are solid, so that their particles cannot move easily, they are drawn to one another without losing their form. It is in this way that we explain the motions of the planets.

We will first consider what form would result from the gravitation of free particles toward each other, and then see how motion would modify this form. We will afterward inquire what motions gravity, acting in connection with a primitive impulse, would impress on these bodies.

§ **329.** The globular form of the sun and planets makes it probable that they were once in a fluid state. A fluid mass, sufficiently removed from other bodies, is always brought to a globular form by the equal mutual attraction of its particles. The fluid particles move over one another with great ease. They continue to move until the attractions are all balanced, and the centre of gravity is at the same distance from every point in the surface of the mass. The only form which answers to this condition is a sphere; for in a sphere the centre of form is also the centre of gravity. The spherical form is accordingly assumed by drops of rain, mist, quicksilver, &c.

If there were no motion in the heavens we should probably find all the heavenly bodies perfect globes or spheres. But we find motion, and this of a kind to affect the form which a body in a fluid state would assume. When a fluid mass rotates, all the parts rotate in the same time. Those which form the axis of rotation may be considered as merely turning round on themselves slowly; those further from the centre describe circles with more rapidity, and those which are farthest from the axis of rotation describe the largest circles, and have the greatest rapidity. By this greater swiftness of rotation in the equatorial parts, centrifugal force is generated, the equatorial particles try to fly off, the equilibrium of the globe (as it would be at rest) is destroyed. A part of the attraction which kept the equatorial particles in their place is balanced by the centrifugal force; the rest is insufficient to retain them in their place; they recede from the centre, and other particles rush in to supply their place.

§ **330.** But the form of the mass is changed; the particles near the equator are piled up, those near the poles are depressed, until it reaches the state of equilibrium for a rotating fluid mass; but if the mass while fluid stops rotating it must return to a globular form. If it hardens while rotating, it will retain the spheroidal form. And its departure from the truly spherical form will be in proportion to the rapidity of rotation. Very rapid rotation causing great centrifugal force raises the equatorial particles and depresses the polar parts exceedingly. The different planets have now varying rapidities of rotation; and those which rotate most rapidly are most flattened at the poles.

It is found by calculation that the rotation of the earth is precisely rapid enough to give it the flattening which it has. If it rotated more rapidly, the water at the equator would be flung off. Or if it were a larger body, and rotated in twenty-four hours as it now does, the equatorial particles would likewise be in danger.

§ **331.** The effect of rotation on the form of a fluid mass may be seen by spinning a pail partly full of water suspended by a string. The surface of the water instead of remaining horizontal will become concave. The centrifugal force generates in all the water a tendency to leave the axis and to press toward the circumference. It is therefore urged against the pail, and forced up the sides, till the excess of height, and consequent increase of pressure downwards, just counterbalances its centrifugal force. If the rotation becomes very rapid, the surface of the water becomes more concave; if allowed to diminish, it becomes less so. In a similar way more or less rapid rotation would increase or diminish the excess of the equatorial parts.

The following pretty little experiment has been tried to show the globular form which a liquid relieved from all external attraction or pressure would take, and also the spheroidal form which rotation would induce.

Placing a mixture of water and alcohol in a glass box, and therein a small quantity of olive-oil, of density precisely equal to the mixture, we have in the latter *a liquid mass relieved from the operation of gravity*, and free to take the exterior form given by the forces which may act upon it. In point of fact, the oil instantly takes a globular form, by virtue of molecular attraction. A vertical axis being introduced through the box, with a small disc upon it, so arranged that its centre is coincident with the centre of the globe of oil, we turn the axis at a slow rate, and thus set the oil sphere in rotation. "We then presently see the sphere *flatten at its poles* and *swell out at its equator*, and we thus realize, on a small scale, an effect which is admitted to have taken place in the planets. The spherifying forces are of different natures, that of molecular attraction in the case of oil, and of universal attraction in that of the planet; but the results are analagous, if not identical. Quickening the rotation makes the figure more oblately spheroidal. When it comes to be so quick as two or three turns in a second, the liquid sphere first takes rapidly its maximum of flattening, then becomes hollow above and below, around the axis of rotation, stretching out continually in a horizontal direction, and finally abandoning the disc, is *transformed into a perfectly regular ring.* At first this remains connected with the disc by a thin pellicle of oil; but on the disc's being stopped, this breaks and disappears, and the ring becomes completely disengaged.

§ **332.** We have spoken as if the fluid mass of the earth were first formed into a globe, and afterward, from rotation, took on the spheroidal shape. Probably the moment it was isolated it took its form, which rotation, beginning at the same moment, modified. The same force which isolated it, probably communicated by one impulse the rotary and onward motion.

The fact that these two motions throughout the solar system almost invariably take place in the same direction, makes the probability that they were communicated by one impulse as millions to one. For simplicity, let us consider the case of the earth only, and study how these two motions might have arisen. Whether the fluid matter which we have good evidence once composed the earth, was brought from some other part of space within the sphere of the sun's attraction, or whether it was thrown off the rotating sun, as splinters are sometimes cast off grindstones, or whether from some other cause it received an impulse, we cannot but suppose it to have had some inclination to move in some direction. Newton supposed that all the planets had received an impulse in a straight line in a tangent to their present orbits. The projectile force could not have passed through the centre of gravity, for the earth rotates. If the direction of the impulse was through the centre of the sphere, equal velocities would have been communicated to all parts of the sphere, and no rotation would have taken place. But if the force was directed a little on one side of the centre of gravity, the equilibrium of the particles would be destroyed. Those which received the blow would be carried down with great rapidity, and the other half of the sphere would rise.

§ **333.** The axis of rotation would immediately be formed in the line where all the opposing forces balance each other, and the particles which received the stroke would move round it, carrying with them all the particles in the same hemisphere with themselves. The place which received the impulse would become a point in the equator.

But simultaneously with the formation of the axis of rotation the sun must have attracted to itself one half of the sphere, at right angles to the direction in which the projectile force acted. A plane passing through the centre of the sun and the line representing the projectile force marks the plane of the ecliptic. If the sun were below the plane of the earth's equator, it would draw the earth down, and cause the ecliptic to cut the equator as it now does. It would make the poles of the orbit inclined as they are to the poles of the equator.

Rotation once established, nothing could occur to change its poles, and we have every evidence that they have remained unaltered as far as man's records extend.

The axis always remains parallel to itself however much it may be inclined to the axis of the onward motion. This may be shown by throwing into the air a homogeneous globe pierced with an axis, and impressing on it at the same time motion of rotation. Whether this axis is perpendicular to the curve described by the body or not, it will always remain parallel to itself.

When the direction of the force does not pass near the centre of gravity, great velocity of rotation is induced. The part of a sphere in which the force has been applied may be found by calculation. In case of the earth it was in some part of the equator, passing about twenty-five miles from the centre. The remoter planets must have been impelled in a direction passing farther from the axis of revolution, for their rotation is extremely rapid.

The rotary motion requires no expenditure of force. It merely results from the unequal application of the force. None of the force applied is used up by it, because the body as a whole is in a state of rest. As many of the particles move backward as forward, and the centre of gravity undisturbed by rotation remains unmoved.

§ **334.** On the earth, where there is so much friction, a blow which is unable to overcome the inertia of a body and friction, will, if its direction passes through the centre, cause the body to shake ; if it passes on one side of the axis it may cause a slight rotary motion. We feel therefore as if force were always consumed in causing rotation. But it is a fact that equal impulses will carry equal bodies over equal spaces in equal times whether the bodies rotate or not. And this is not only true of a rotating body, but, as we shall presently see, of a system of bodies revolving round a common centre. On the earth, when friction is removed as much as possible, a very slight force is sufficient to destroy equilibrium and cause motion. If a hundred pounds could be placed in the scale of a delicate balance, a weight scarcely more than sufficient to overcome the friction of the machine would cause one scale to descend and the other scale to ascend. In the heavens, where there is no friction to be overcome, the slightest inequality of the two forces acting on the balanced particles of a sphere is sufficient to make the whole sphere rotate.

§ **335.** The onward motion, or translation of a body always requires an expenditure of force, and equal forces will carry the centre of gravity forward through equal spaces, in whatever part they may be applied. This is true of different parts of a solid body ; of two bodies connected by a pole ; or of two unconnected bodies.

Suppose two bodies, weighing one pound each, and connected by a pole, to receive through their centre of gravity a blow which carried them forward two feet. An equal blow, not on the centre of gravity of the two bodies, but on one of them, would, if that were alone, carry it forward four feet ; but as it is fastened to the other body, the force is divided between the two, and therefore carries them and their common centre of gravity forward two feet ;

producing also in the two bodies a whirling motion round their centre of gravity. If the bodies have no connection, and one receives a blow, it moves forward four feet, but the other is stationary; the centre of gravity is half way between the two, and in this case also advances two feet.

A cluster of stars may thus move onward unimpeded by their individual revolutions round their common centre of gravity, and the earth and moon make their little evolutions, while their common centre of gravity sweeps steadily round the sun.

§ **336.** Since the common direction of the two planetary motions points to a common origin, and since one impulse can account for both motions we need seek no other cause for either motion. Not only is it very probable that one cause produced both motions, but the chances are millions against one that any cause should have produced one alone. So that if we find rotation or revolution alone we should be inclined to suppose that both motions had existed, and that one had been stopped by some external force. To cause revolution alone, we have seen that the direction of the force must be through the centre, which is extremely improbable. Rotation alone can be caused on earth when onward motion is destroyed by friction, but in the heavens, where there is no resistance, any force whose inequalities could cause rotation, must itself cause revolution. Wherever there is revolution, therefore, rotation is probable, as in the case of the planets most distant from the sun, which have not been observed. Wherever there is rotation we may infer translation in space, even when we cannot perceive it, as in the case of the sun. Or we may suppose that a force acting through the centre of gravity in the opposite direction, has put an end to the onward movement without interfering with rotation. The two motions are independent of one another, and one may outlast the other. Thus a top often twirls on the same spot after it has ceased to describe circles.

The period of a planet's rotation depends on the place in which it is struck, and the force of the blow compared to its mass.

The period of a planet's revolution depends on the mass

and distance of the attracting body and the force of its projection into space.

§ **337.** In consequence of rotation every particle on the surface of the earth describes a circle once in twenty-four hours. In consequence of revolution, the whole earth, and of course each place on its surface, describes, in the course of 365 days, an ellipse, of which the sun is in one of the foci. We have learned the influence of gravity on the shape of bodies, we will now seek the effect of gravity on a system of bodies, in establishing a common centre, and consider the motions which gravity and a primitive impulse would impress on bodies.

The simple action of gravity can only be seen in the heavens, in the attraction one heavenly body exerts on another. When bodies on the surface of the earth are drawn to the earth, this drawing, which we call their weight, is not the full measure of gravity. Other forces are at work. The whirling of the earth gives all bodies in inhabitable parts of its surface a tendency to fly off, which partly counteracts the force of gravity. The gravitation of a body to the earth is then the force with which gravity draws it, minus its centrifugal force. Even at the equator, however, the difference between the power of gravity and the actual gravitation is trifling; it is but $\frac{1}{289}$ of the whole weight. As the tendency to fly off can be calculated, the true amount of gravity at the earth's surface can be ascertained. At the poles the gravitation of bodies is an exact measure of the force of gravity residing in a body of the mass of the earth, and exercised on a body removed from the centre of gravity by a distance equal to the polar radius.

§ **338.** The first law of gravitation is this. The attraction of one body on another does not depend on the mass of the body attracted, but is the same whatever be the mass, if the distances are the same.

Thus Jupiter attracts the sun, and Jupiter attracts the earth also; but though the sun's mass is three hundred thousand times as great as the earth's, yet the attraction of Jupiter on the sun is exactly equal to his attraction on the earth, when the sun and the earth are equally distant from Jupiter. When the sun and earth are at equal distances from Jupiter, the attraction of Jupiter on the sun draws it through as many inches or parts of an inch, in one second of time, as it draws the earth in the same time.

The second law of gravitation is this. Attraction is proportional to the mass of the body which attracts, if the distances of different attracting bodies are the same.

Thus suppose the sun and Jupiter are at equal distances from Saturn. The sun is about 1,000 times as large as Jupiter. Then whatever be the number of inches through which Jupiter draws Saturn in one second of time, the sun draws Saturn in the same time through 1,000 times that number of inches.

The third law is this. If the same attracting body acts upon several bodies at different distances, the attractions are inversely proportional to the squares of the distances from the attracting body. Thus the earth attracts the sun, and the earth also attracts the moon. But the sun is 400 times as far off as the moon, and therefore the earth's attraction on the sun is only $\frac{1}{160000}$ part of its attraction on the moon. Or as the earth's attraction draws the moon through about $\frac{1}{20}$ of an inch in one second of time, the earth's attraction draws the sun through $\frac{1}{3200000}$ of an inch in one second of time.

§ **339.** The reader may ask, "How is all this known to be true ?" The best answer is perhaps the following. We find that the force which the earth exerts upon the moon bears the same proportion to gravity on the earth's surface which it ought to bear in conformity with the rule just given. For the motions of the planets, calculations are made which are founded upon themselves, and which will enable us to predict their places with considerable accuracy if the laws are true, but which would be much in error if the laws were false. The accuracy of astronomical observations is carried to a degree which can scarcely be imagined. And by means of these we can every day compare the observed place of a planet with the place which was calculated beforehand, according to the law of gravitation. It is found that they agree so nearly as to leave no doubt of the truth of the law. The motion of

Jupiter, for instance, is so perfectly calculated, that astronomers have computed ten years beforehand the time at which it will pass the meridian of different places, and the predicted time is found to be correct within half a second of time.

§ **340.** Since all the planetary bodies are of a spheroidal form, the labors of astronomers are much shortened by considering all lines of attraction as passing from one centre to another. In a sphere the centre of form is also the centre of gravity, that is, it is the point round which the weight of the body is equally distributed in all directions.

It is often necessary to ascertain the position of the common centre of gravity of two or more bodies connected together; and this is not difficult when we know the centres of gravity of the several bodies themselves. (Fig. 6, Plate I.) Let A and B be two globes connected by a rod, which we shall suppose, for the sake of simplifying the explanation, to have no weight in itself. The centres of these globes are their centres of gravity. And as we may regard all their weight as acting from those points, the same reasoning which enabled us to understand that the forces acting upon the different parts of any one body may balance each other round a certain point, leads to the belief that a point may exist in which we may regard the actions of A and B as jointly and equally exercised. This point is evidently somewhere in the line A B, which joins their centres. It is determined on the principle of the lever, by dividing the line A B into two such parts that the distance of each body from the point C shall be proportional to the weight of the other. Thus suppose A to weigh 6 lbs., and B 1 lb., then A's distance from C must be to B's distance as one to six. A support placed at C will sustain them both at rest, and will be pressed upon with the weight of both combined. But owing to the greater distance of B from C, and its inferior size, a given force applied to it will carry it through six times the space through which the same force would carry A. If both bodies therefore were moving round their common centre of gravity, they would perform their orbits in the same time, but B would move with six times the velocity of A.

§ **341.** Supposing that a third body, D, were connected with A and B, by a rod proceeding from the point C. Then the common centre of gravity of all three bodies will be in the line C D, since the weights of A and B may be regarded as concentrated at C and act as if a single body of their total weight were placed there ; and it may be determined in the same manner as before. In like manner, if another body, F, be connected with the system, by a rod uniting it with the rest at their common centre of gravity, E, the centre of gravity of the four will be in the line between E and F. And it will be at such a distance from F, that its weight multiplied by the distance F G, shall be equal to the combined weights of the other bodies acting at the distance G E. In this manner any number of bodies may be connected with the system ; or in a system already existing, we may ascertain the common centre of gravity by a similar process.

As the rod connects the globes so does gravity hold the celestial bodies in their places. But since all, at least in our system, are in motion, the balance must be each moment struck anew. Each orb is exposed to the influence of every other orb, and by their incessant deviations from regular motion, the equilibrium and stability of the whole are preserved. When Jupiter passes on the same side of the sun with the earth, the earth cannot but feel his attraction, and Jupiter and all his moons recognize the approach of the little earth. A balance is struck between the sun's attraction and that which Jupiter in his present place exercises on the earth. The earth moves toward Jupiter till the equilibrium is restored and modifies her path by his influence. All this is not done by jerks or intervals. Gravity acts incessantly, restoring order as rapidly as it is disturbed. Every change is effected gently, swiftly, and so far as we can judge noiselessly.

§ **342.** Let us suppose two bodies newly suspended in space ; both bodies would rush together and meet at one point. If one body were heavier than the other, the point of meeting would be proportionally near the heavier body.

If they were equal in weight, they would meet mid-way. Now let us suppose the second body to have an onward motion given to it. The moment it felt the attraction of the first body, it would attract that in return, share its motion with it, and force it to move round their common centre of gravity. If the two bodies were of equal weight, they would revolve at equal distances from the centre of gravity; if not, the heaviest would move in the smallest orbit; but both would revolve in the same time.

CHAPTER XV.

LAWS OF MOTION—(CONTINUED.)

Three general Laws of Motion. Composition of Forces. Path of a Projectile near the Earth's Surface. Motion in a Curve. Projectile and Centripetal Forces. Motion in the Solar System. Kepler's Laws. Central Forces.

§ **343.** There are three general laws which a body obeys in its motion, whatever be the kind of body or the kind of force that impels it; whether it be a particle of dust driven by the wind, or a planet revolving in consequence of an original impulse through the celestial spaces.

A body does not change its state, either of rest or motion, unless in consequence of some external cause.

The effect is always proportional to the force impressed, and takes place in the direction in which the force acts.

Action and reaction are equal and contrary. This law holds whether the bodies attract or repel one another, and whether they act at a distance or in apparent contact.

A body is often acted on by several forces at once, and the effect of their joint action is an exact compound of their several effects, or the same as if each had acted successively.

The body may be acted on by two forces applied at the same point of the body. If they act in the same direction, the resultant will be in that direction, and equal to their sum. If in opposite directions, the resultant will be in the direction of the greater, and equal to their difference. If at an angle, the resultant will be in the same plane, and represented by the diagonal of a parallelogram of which the two sides represent the simple forces. In this case it is less than the sum of the forces, and greater than their difference.

Thus a body may move in a certain direction in consequence of one, two, or a dozen impulses.

It is often desirable to resolve a single force into others to which it is equivalent, in order to find its effect in a given direction. We need only resolve the force into two, one of which is parallel to the given direction, and the other at right angles to it. The latter can have no effect in the given direction, and therefore the other will express the whole effect.

Thus when we wish to know how much one planet draws another from the plane of its orbit, the line representing the influence of the planet and in the direction of their centres is considered the diagonal of a parallelogram of which one side shows how much the planet is moved from its orbit.

Celestial motions are caused, not by two impulses, but by an impulse and a pressure. The impulse was imparted by a force of whose nature we can only form a vague guess. The pressure is the constant attraction of gravitation. If the heavenly bodies had received only the impulse, they would have moved on in straight lines forever. If other impulses had interfered successively, they would have moved on in broken lines, making angles with each other. Since the second force is a pressure, and acts incessantly, the straight lines will become infinitely short, and the path consequently will become curvilineal. Since a body, if left to itself, moves in a straight line, we may conclude, when it moves in a curve, without being compelled to it by a fixed obstacle, that there is a force of pressure constantly deflecting it from the direction of the tangent. We are now therefore ready to consider the effect of attraction on the motion of bodies.

§ **344.** If a body is projected in the direction in which gravity draws it, its velocity is increased. If gravity acts directly contrary to the projectile force, it gradually weakens and at length overcomes it, as when an arrow shot vertically is brought to the ground. A more important case for astronomy than either of these is when the body is projected transverse to the direction in which the force draws it.

The simplest instance of this motion that we can imagine is the motion of a stone when it is thrown from the hand in a horizontal direction. It does not move in a straight line. It begins to move in the direction in which it is thrown; but this direction is speedily changed. It continues to change gradually and constantly, and the stone strikes the ground moving at that time in a direction much inclined to the original direction. The most powerful effort that we can make is not sufficient to prevent the body from falling at last. This experiment therefore will not enable us to judge immediately what will become of a body (as a planet) which is put in motion at a great distance from another body which attracts it (as the sun). But it will assist us much in judging generally what is the nature of the motion when a body is projected in a direction transverse to the direction in which the force acts upon it.

§ **345.** The general nature of the motion is this. The body describes a curved path, of which the first part has the same direction as the line in which it is projected.

If A (Fig. 8, Plate I.) is the point from which the stone was thrown, and A H the direction in which it was thrown; and if we wish to know where the stone will be at the end of any particular time, (suppose three seconds,) and if the velocity with which it is thrown would, in three seconds, have carried it from P to F, supposing gravity not to have acted upon it; and if gravity would have made it fall from A to P, supposing it to have been merely dropped from the hand; then, at the end of three seconds, the stone really will be at the point F. And it will have reached it by a curved path A F, of which different points can be determined in the same way for different instants of time.

The calculation of the stone's course is easy, because during the whole motion of the stone gravity is acting upon it, with the same force and in the same direction. The motion of a body attracted by a planet or the sun, where the force varies as the distance alters, and is not the same either in amount or direction at the point F as it is at the point A, cannot be computed by the same simple method. But the same method will apply, provided we restrict the intervals for which the calculations are made to times so short that the alterations in the amount of the force and in its direction, during each of those times will be very small. Thus, in the motion of the earth, as affected by the attraction of the sun, if we used the process that we have described to find where the earth will be at the end of a month from the present time, the place that we should find would be very far wrong. If we calculated for the end of a week, since the direction of the force and its magnitude would have been less altered, the error would be much less than before.

Fig. 8, Plate I., shows the paths described in obedience to several different combinations of the projectile force with gravity.

§ **346.** Every body which is under the influence of a constant attractive force, and of an impulse originally given, moves in a curved path round the centre of attraction.

And on the other hand, when a body moves in a curve, there must be one impulsive and one constantly restraining force acting upon it. If the projectile force acted at perceptible intervals, the body would describe the perimeter of a polygon, having as many sides as the number of impulses given. Since the projectile force acts at intervals infinitely small, the polygon has an infinite number of sides, is a circle. The projectile force urging the body on at a tangent, diminishes its tendency to the centre or the centripetal force, and creates a centrifugal force in the opposite direction from the centripetal force. The centrifugal force arises from and is inseparable from a curvilinear track. It is not a tendency which the body originally has to fly from the centre, but arises from its constrained continuance in a curved orbit, when, if unattracted to the centre, it would proceed in a tangent.

The projectile and centripetal forces cannot be directly compared, for they are different in kind, one is impulsive and the other is incessant; but the centrifugal force, generated by a given projectile force, acts incessantly like the centripetal, and may be compared with it. We may also calculate through how long a space gravity must act to balance a given projectile force.

§ **347.** Every body moving under the influence of gravity describes one of the conic sections.

If a cone is cut parallel to its base the section is a circle; if cut obliquely to its base, but not in such a way as to intersect it, the section will be an ellipse more or less elongated. If it is cut parallel to the curved surface of the cone, the section will be a parabola; if perpendicular to the base and not through the axis of the cone, it will be a hyperbola. A cone divided by these sections best explains this. The curve described cannot be a circle unless the line of projection is perpendicular to the line of attraction, and unless the velocity with which the planet is projected is neither greater nor less than one particular velocity determined by the distance and mass of the attracting body. If it exceeds this velocity a little, or falls a little short of it, the body will move in an ellipse.

If the projectile force gives a rapidity equal to that acquired by a body in falling through a height equal to one third the radius of the circle, and if also the projectile act at right angles to the centripetal force, the body will describe a circle.

If the projectile force is to that required for a circle as $\sqrt{2}$ to 1, or equal to that acquired by a body falling through one half of the radius, the body will describe a parabola.

Any ratio of the central forces between these two will cause the body to describe an ellipse.

If the projectile force is stronger than that required for a parabola, or such as a body would gain by falling through a height yet greater than one half of the radius, the orbit will be hyperbolical.

§ **348.** As it is extremely improbable that the two forces should be to one another in the definite proportions required to cause a circle or a parabola, we cannot expect to find these orbits in the heavens.

Since ellipses and hyperbolas require no definite ratio of the forces, but vary in their axes according to the ratio, we may expect to find many orbits of these forms. Indeed we have no certainty that any except elliptic orbits exist in the heavens. The comets which have yet been recorded but once, may thousands of years hence manifest their elliptic orbits by revisiting us again. Meanwhile it requires a nice observation and calculation, to ascertain from the small portion of their orbit visible to us, whether it most resembles an ellipse, parabola, or hyperbola.

If the projectile acts obliquely to the attracting force, and the velocity of projection is small, the body will move in an ellipse. If the velocity is great, it may move in a parabola or a hyperbola, but not in a circle. For even if the velocity of projection were just sufficient to make the body move in a circle, yet as it acts obliquely to gravity, gravity must either diminish or increase it, and thus prevent the body's moving in a circle.

The earth when nearest the sun has a velocity of about 102,300 feet a second, this being the result of its own projectile force, arising from the impulse which first set it in motion, and the power of the sun's attraction. By reason of this velocity it is constrained to move in an elliptic orbit. But if by any augmentation of the projectile force, the earth's velocity at this point were to amount to 144,700 feet, the orbit would become parabolic; and any velocity surpassing this would make its course hyperbolic. If its velocity were about 101,000 feet a second, or a little less than it now is, the orbit would be exactly circular.

Thus various forms of orbits are produced by different projectile forces combined with one given attractive force. As the projectile force increases the body departs more and more from a circle.

If a body describes a circle, the attracting body is in the centre of the circle.

If it describes an ellipse, the attracting body is not in the centre of the ellipse, but in one focus.

If it describes a parabola or hyperbola, the attracting body is in the focus.

§ **349.** The student should familiarize himself with the following terms which occur continually in speaking of ellipses.

A straight line drawn from any point of the curve to the centre of attraction is called the *radius vector*.

The *angular velocity* at any point of the curve is the velocity with which the radius vector at that point describes an angle. The actual velocity is equal to the space passed over divided by the time. Two bodies revolving round a common centre may have the same angular velocity; but if one be twice as far from the centre as the other, its actual velocity will be twice as great.

When the line which the body describes returns into itself, like a circle or an oval, it is called an *orbit*, and the time of describing the whole is called the *periodic time*.

Fig. 9, Plate I. In the ellipse A E B D, S and H are the *foci*. Let S be that focus which is the place of the sun, if we are speaking of a planet's orbit, or the place of the planet, if we are speaking of a satellite's orbit.

A B is the *major axis* of the ellipse.

A C or C B is the *semi-major axis*. This is equal in length to S D. It is sometimes called the *mean distance*, because it is half way between A S, which is the planet's smallest distance from S, and B S, which is the planet's greatest distance from S.

D E is the *minor axis ;* D C or C E the *semi-minor axis*.

A is called the *perihelion*, B the *aphelion* of the orbit of a planet. In the moon's orbit they are called the *perigee* and *apogee*.

A and B are called the *apsides*, and the major axis *the line of apsides*.

The proportion which S C bears to A C is called the *excentricity of the orbit*.

§ **350.** If we know the mass of the central body, and if we suppose the revolving body to be projected at a certain place in a known direction with a given velocity, the length of the axis major the eccentricity, the position of the line of *apsides*, and the periodic time may all be calculated.

In all our diagrams it is to be understood, that the planet, or satellite, moves through its orbit in the direction opposite to the motion of the hands of a watch. This is the direction in which all the planets and satellites would appear to move, if viewed from any place on the north side of the planes of their orbits.

The deflection of a stone, thrown from the hand, from the straight line in which it began to move, exactly equals the space through which gravity would have made it fall in the same time from a state of rest, whatever may be the velocity with which it is thrown. Consequently when the stone is thrown with very great velocity, it will go a great distance before it is much deflected from the straight line, and therefore its path will be very little curved.

The same thing is true with regard to the motion of a planet. And thus the curvature of any part of the orbit which a planet describes will not depend simply on the force of the sun's attraction, but also on the velocity with which the planet is moving. The nearer the planet is to the sun, the more it will be drawn in, and its orbit curved; but at the same time the greater velocity of the planet at any point of its orbit will tend to diminish the curvature of the orbit at that part. The same absolute curvature may be produced, as at the upper and the lower end of an ellipse, by two more or less powerful centripetal and centrifugal forces.

§ **351.** Let us follow the planet through every part of its orbit. Suppose it projected at B with the necessary velocity. This velocity must not be so great that the attraction of the sun will not bend its path very much. From B to D and A the attraction exceeds the centrifugal force, and velocity towards the centre is created. The sun's attraction increases this velocity as the body moves towards A. The sun's attractive force also, on account of the planet's nearness, is very much increased at A, and tends to make the orbit more curved; but the velocity is so much increased that the orbit is not more curved than at B. At

A the attractive and centrifugal forces are equal, and from A to E and B the centrifugal exceeds the attractive, so that between A and B the velocity towards the centre is destroyed. After B the attraction is again in excess, and between that and A the velocity to the centre is increased. The sun's attraction retards it as the force of gravity retards a ball which is rolled up hill. When it has reached A its velocity is comparatively small; and therefore, though the sun's attraction at A is small, yet the deflection which it produces in the planet's motion is, on account of the planet's slowness there, sufficient to make its path very much curved, and the planet approaches the sun, and goes on the same orbit as before.

§ **352.** The projectile force is that with which the circulating body would run off in a tangent to its path, if there were no centripetal force to prevent it. The projectile force at the first moment the body begins to describe a curve depends on the strength of the initial impulse. But after gravity has begun to act, the projectile force is increased or diminished according as gravity acts in or against the direction in which the body is moving. It must be remembered that it is the tendency which the body has to fly off in a tangent, and the strength of this tendency depends on the velocity of the body at each moment. This velocity may be greater or may be less than the initial velocity. When the body describes an ellipse, the projectile force is alternately greater and less than the initial velocity. When it describes a circle the velocity and the projectile force are the same. When a body is made to describe a circle, by moving along the concave surface of a sphere or cylinder, however much the velocity may increase the centrifugal force, the reaction of the surface will be increased in the same proportion, so that the body may describe the same circle with different degrees of velocity.

But if the body is moving through space round a centre, and a great velocity be given to it, the centrifugal force will exceed the centripetal. Therefore the body will be driven to a greater distance than before, from the centre, and will describe a curve exterior to the circle. For a like reason, if the velocity be diminished, the centrifugal force becoming less than the centripetal, the body will describe a curve interior to the circle.

In general, the centrifugal force is directly as the square of the velocity of the revolving body, and inversely as the radius of curvature of the arc which it describes.

§ **353.** Three important laws of planetary motion were discovered by Kepler, and pass under his name.

I. The orbits of all the planets are ellipses, of which the sun occupies one focus.

II. The radius vector of the planet describes equal areas in equal times.

III. The squares of the periodic times are as the cubes of the mean distances of the planets, or as the cubes of the major axes of their orbits.

The fact of the ellipticity of the orbits he determined from direct observation of Mars when in different parts of his orbit. Finding the distances of Mars incompatible with the supposition that he moved in a circle, he tried them with an ellipse, and found they corresponded. Upon trial of the other orbits, he found them also elliptic.

The variations in the apparent diameter of the sun likewise agree with the supposition that the orbit is elliptic.

If we draw an ellipse, and from one focus, the position of the sun, draw radii to the positions of the earth at equal intervals, thus dividing the surface of the ellipse into sections, these sections will be equivalent. Those which rest on short arcs are included within longer sides, and those which have long arcs are included within shorter sides. Thus the radius vector describes equal areas in equal times.

Having discovered the relative mean distances of the planets from the sun, and knowing their periodic times, Kepler endeavored to find if there was any relation between them, and thus discovered his third law.

§ **354.** The following are the most important propositions concerning the motion of bodies acted on by central forces.

If the centre of attraction remains always in the same place, the curve will be wholly in one plane, passing through that centre; and the areas described by the radius vector will be proportional to the times of description.

First. Suppose the central force to act during equal finite intervals of time; suppose C (Fig. 1, Plate II.) the centre of attraction, A B the line passed over in one of the equal intervals, the body with its uniform motion would, during the next equal portion of time, go over a line B D =A B, but at B it is acted on by the central force. Suppose the momentary action is such, that in the same time the body would move along B E, then completing the parallelogram, B F will be the real line of motion. Joining C D, the triangle A B C=triangle B C F, because they have equal bases, and the same altitude. The same thing may be shown with regard to the next triangle, &c. Hence, the sum of all the triangles, or the whole area described in a given time, will be proportional to the time of description.

As the force acting on the body is supposed to be in the direction of the plane of A B C, it has no tendency to move it out of that plane, and therefore B C F is in the same plane with A B C. The same is true of the next triangle, and of the whole area described.

§ **355.** If the curve described lies wholly in one plane, and the radius vector, drawn from a certain point in the plane, always describes around that point areas proportional to the times, that point is the centre of attraction.

For around any other point than the centre of attraction, the areas described in equal times cannot be equal. Thus, (Fig. 1, Plate II.,) take any point, G, it is evident that G B F cannot be equal to G B A; for then D F would need to be parallel to G B, whereas it is parallel to C B.

§ **356.** The projectile velocity at any point of the curve is inversely as the perpendicular let fall on the tangent at that point from the centre of attraction.

For the small triangle described in an instant, by the radius vector being every where of the same area, its base must be inversely as its perpendicular; but the base is the projectile velocity, and the perpendicular on the base is the perpendicular on the tangent.

§ **357.** If there be two free bodies, the one cannot remain at rest, while, by its attraction, it causes the other to move round it; but if the two bodies receive equal impulses in opposite and parallel directions, their centre of gravity will remain at rest, and they will describe similar curves. The first part of the proposition is manifest, for as one body attracts the other, the other will attract the first, and cause it to approach.

Let A and B (Fig. 2, Plate II.) be the two bodies, C their centre of gravity. It follows, that A C : C B : : B : A. Let the bodies receive equal and parallel impulses in the directions B F and A G, and suppose that, in a given interval, the body A would move along A G; join G C, and produce it to F; B F will be the line passed over by B in the same moment; for the impulses being equal, the velocities will be inversely as the masses, that is, directly as A C : B C; but by similar triangles A G : B F : : A C : B C. Again, suppose that, in consequence of the mutual attraction, the bodies describe the curves B D, A E, then G E, F D will be the momentary deflections. G E will be to F D : : B : A; G C : C F : : A C : C B : : B : A; and hence the remainder E C will be to the remainder C D also in the same proportion, viz :—: : B : A. Hence the same point C will still be the centre of gravity of the two bodies when they have arrived at E and D, and will be so continually.

Again, the small arcs A E and B D are similar, since all the straight lines connected with the one are proportional to the corresponding lines connected with the other. The arcs described the next moment will be similar for a like reason; and hence the whole arcs described in equal finite times will be similar, and the whole curves described by A and B will be similar.

§ **358.** The angular velocity at the centre of force is inversely as the square of the radius vector. For the area of the indefinitely small triangle A B C (Fig. 1, Plate II.) is expressed by $BC^2 \times$ angle A C B. But in equal intervals of time, these areas are equal in all parts of the orbit. The angle A C B must therefore be inversely proportional to BC^2, or to the square of the radius vector.

§ **359.** To determine the ratio of forces by which bodies tending to the centres of given circles are made to revolve in their peripheries. Let A M a (Fig. 3, Plate II.) be the circle in which one of the bodies moves round the centre of force E, and let the indefinitely small arch A O be the distance it moves over in a given small interval of time. The centripetal force will be proportioned to A p. The chord and arc A O will be equal in length. Whence, $AO^2 = Aa \times Ap = AC \times 2\,ap$; consequently, $2\,Ap = \frac{AO^2}{AC}$. And the same may be shown with respect to motion in any other circle. So that if R and r denote the radii of two circles, F and f the respective central forces, V and v the velocities with which the bodies move in their peripheries, we shall have $F : f :: \frac{V^2}{R} : \frac{v^2}{r}$; therefore the forces are as the squares of the velocities directly, and as the radii inversely.

Cor. 1st. Because $F : f :: \frac{V^2}{R} : \frac{v^2}{r}$, it follows that

$$V : v :: \sqrt{RF} : \sqrt{rf}, \text{ and}$$

$$R : r :: \frac{V^2}{F} : \frac{v^2}{f}.$$

§ **360.** The centrifugal force may now be compared with gravity, for if v be the velocity of a particle moving in the circumference of a circle of which r is the radius, its centrifugal force is $f = \frac{v^2}{r}$. Let g be the constant force of gravity, and h the space or height through which a body must fall in order to acquire a velocity equal to v; then $v^2 = 2\,hg$. for the accelerating force in the present case is gravity; hence $f = \frac{2\,hg}{r}$. If we suppose $h = \frac{1}{2}r$, the centrifugal force becomes equal to gravity.

Thus, if a heavy body be attached to one extremity of a thread, and if it be made to revolve in a horizontal plane round the other extremity of the thread fixed to a point in the plane; if the velocity of revolution be equal to what the body would acquire by falling through a space equal to half the length of the thread, the body will stretch the thread with the same force as if it hung vertically.

§ **361.** Since the versed sine of an arc of a circle is equal to the square of the corresponding chord divided by the diameter, and the chord of a very small arc nearly equals this arc, the square of this arc divided by the diameter, gives the versed sine. Let r be the radius, π the ratio of the circumference of a circle to its diameter, and T the time of revolution expressed in seconds. The arc actually described in a minute of time $= \frac{2 r \pi}{T}$. This squared and divided by the diameter gives, for the versed sine, which is proportional to the attracting force, the expression $\frac{2 r \pi^2}{T^2}$. But the attracting forces are in inverse proportion to the squares of the radii. Hence we have

$$\frac{1}{r^2} : \frac{1}{r'^2} : : \frac{2 r \pi^2}{T^2} : \frac{2 r' \pi^2}{T'^2}$$

$$r^3 T'^2 = r'^3 T^2 ;$$

which is Kepler's third law, that the squares of the periodic times are proportional to the cubes of the mean distances.

This same proposition gives us the mass of two attracting bodies, the orbits and periodic times of two bodies revolving round them being known. The forces are the masses, and may be found by dividing the cube of the distance of a body moving round one of them divided by the square of its time, with the cube of the distance of the other divided by the square of its periodic time. Thus the cube of the moon's distance divided by the square of its periodic time is to the cube of the earth's distance divided by the square of her periodic time (nearly) as the mass of the earth is to that of the sun.

§ **362.** By means of the preceding propositions several practical questions of interest are solved. The distance of a revolving body from the earth's centre being known, its velocity and periodic time may be deduced. Thus, let the radius of the earth (=21,000,000 feet, nearly,) be denoted by r, and the space through which a heavy body falls in one second at the surface ($=16\frac{1}{2}$ feet), by $\frac{1}{2}$g, the force of gravity at the surface being denoted by g; then will the velocity per second in a circle at the surface be$=\sqrt{2\,g\,r}=26{,}000$ feet nearly; and the time of revolution $=5{,}075$ seconds. Let R be put for the radius of any other circle described by a projectile about the earth's centre: then, because the force of gravitation about the surface varies inversely as the square of the distance, we have, $\frac{1}{\sqrt{r}} : \frac{1}{\sqrt{R}} :: 26{,}000$ feet (velocity per second at the surface) : $26{,}000\sqrt{\frac{r}{R}}$, the velocity in the circle whose radius is R. And $r^{\frac{3}{2}} : R^{\frac{3}{2}} :: 5{,}075s$ (the periodic time at the surface) : $5{,}075\sqrt{\frac{R^3}{r^3}}$, the periodic time in the circle whose radius is R.

For example, if R be assumed equal to 60 r, the distance of the moon from the earth, the expression for the velocity will become $3{,}356\frac{1}{2}$ feet per second; and that for the periodic time will become 2,360,035s. or $27\frac{3}{10}$ days, nearly.

§ **363.** Or, knowing the periodic time of the moon, and the radius of its orbit (240,000 miles), we can calculate the space through which she would fall if left to herself, in a minute. This space will be the versed sine of the arc described in that time. The arc is easily found; for as the moon takes 27 days, 7 hours, 43 minutes, to describe her whole orbit, the following proportion will give it. As $27, 7', 43'' : 1'' :: 360° : 33''$ nearly; of this arc the versed sine may be computed; it is one half of a tenth of an inch, taking the moon's distance to be 240,000 miles.

Now if the force which retains the moon in her orbit is identical with terrestrial gravity, it must decrease as the square of the distance. That is, if the moon is 60 times as far from the earth's centre as a body near the surface of the earth, the space described by the moon should be to that described by a falling body near the earth's surface as $60^2 : 1^2$; the time being the same. A body falls through 59,400 feet in one minute near the earth's surface; hence, as $60^2 : 1^2$; or as 3,600 : 1 : : 59,400 feet : $16\frac{7}{9}$ feet, which is the space through which a body would fall in a minute at the distance of the moon. Now this agrees, making allowance for using round numbers, with the actual distance through which the moon would fall if the centrifugal force were to cease. The moon therefore is retained in her orbit by gravity, and gravity only, for it would be unphilosophical to assign two causes to account for effects precisely similar.

§ **364.** Thus also the ratio of the forces of gravitation of the moon towards the sun and the earth may be estimated. For, $365\frac{1}{4}$ days being the periodic time of the earth and moon about the sun, and 27.3 days the periodic time of the moon about the earth; also 60 being the distance of the moon from the earth in terms of the earth's radius, and 23,920 her mean distance from the sun in the same measure, we have $\frac{23920}{365.25^2} : \frac{60}{27.3^2} :: F : f :: 2\frac{2}{9} : 1$ nearly; that is, the moon's gravitation towards the sun is to her gravitation towards the earth as $2\frac{2}{9}$ to 1 nearly.

Again, from the same principles, the centrifugal force of a body at the equator, arising from the rotation of the earth, is derived. For these propositions apply to centrifugal forces as well as centripetal ones, the terms, being correlatives (where these two alone keep the body in its orbit). And we have just found that the time of revolution is 5,075s. when the centrifugal force becomes equal to the gravity; also it appears that the forces in circles having the same radii are reciprocally as the squares of the periodic times; hence, therefore, since the earth's rotation is performed in 23h. 56m., or 86,160s., we have $86{,}160^2 : 5{,}075^2$: : the force of gravity : the centrifugal force of a body at the equator arising from the earth's rotation : : 1 : $\frac{1}{289}$ nearly.

§ **365.** Since the time of revolution of a body under the equator, and in any parallel of latitude, is equal; the centrifugal forces are as the distances from the axis of motion, or, as the radius to the cosine of the latitude. But in any latitude the centrifugal force is not (as under the equator) opposite to the whole gravity, but only a part of it; which also is to the whole as the cosine of the latitude to radius.

Therefore combining these two ratios, it follows, that the diminution of gravity at the surface of the earth, arising from the centrifugal force, varies as the square of the cosine of the latitude.

The law just stated for the diminution of gravity is on the supposition of the earth's sphericity; but as the polar axis of the earth is rather shorter than the equatorial, the former being to the latter nearly as 300 to 301, or what is technically denominated the compression being about $\frac{1}{300}$, the preceding theory is not exact.

§ **366.** The pendulum serves as an excellent measure of the force of gravity, for by it may be ascertained the distance through which a body unsupported would fall in a second of time. The following proportion will always give that space.

As 3.1416 : 1 : : 1 second : the time of falling through a space equal to half the length of a pendulum beating seconds.

But the spaces described by falling bodies are in the proportion of the squares of the times; therefore:

As the square of the time last found is to one second squared, so is half the length of the pendulum beating seconds to the space through which a body would fall in a second; but this is the measure of the force exerted by the attraction of gravity.

For example, suppose it is found by observation that the length of a pendulum beating seconds, in the latitude of London, is 39.126 inches, or 3.2605 feet; required the space through which a body would fall in a second in that latitude.

As 3.1416 : 1 : : 1 sec. : 3183 sec., the time a body would take to fall through 1.63025 feet, half the length of the pendulum beating seconds.

Again: As 3183^2, or 10131489 sec. : 1^2 sec. : : 1.63025 feet, one half the length of such a pendulum, in latitude $51\frac{1}{2}°$: 16.09 feet, the space through which a body would fall in a second, in that latitude.

§ **367.** Suppose the shorter axis of an ellipse to diminish continually, the longer axis remaining the same; the ellipse will be transformed into a straight line, equal in length to the major axis. In all the successive ellipses produced by this gradual diminution of the minor axis, the periodic time remains unchanged, if the force acting at the centre remains unchanged. The ellipse may be considered as undistinguishable from the major axis, and the revolution in such an ellipse as undistinguishable from the ascent of a body along the axis, to its subsequent descent in an equal time. Consequently a body solicited by such a central force will descend through the space in half the time of revolution in the ellipse. Let T be the time of revolution of a planet at any distance, and t the time of a revolution at half that distance; then, by the third law of Kepler, $T^2 : t^2 :: 2^3 : 1^3$; hence, $t = \frac{T}{\sqrt{8}}$, and $\frac{1}{2}t = \frac{T}{\sqrt{32}}$; but we have just seen that $\frac{1}{2}t$ is the time in which a body would fall to the sun from the distance corresponding to T; therefore, the time in which a planet would fall to the sun by the action of the centripetal force is equal to its periodic time divided by $\sqrt{32}$; or it is equal to that time multiplied by the reciprocal of $\sqrt{32}$, that is, by 0, 176776. By this general rule, the times in which the different planets would reach the sun, if let fall when at their mean distances, may be determined.

CHAPTER XVI.

PERTURBATIONS.*

Disturbing Forces. Problem of the three bodies. Stability of the Solar System. Periodical and Secular Inequalities. Perturbations in Longitude. Motion of the Line of Apsides. Variation of the Eccentricities. Perturbations in Latitude. Retrogradation of the Nodes. Variation of the Inclinations. Permanency of the Major Axes. Effect of a Resisting Medium. Invariable Plane of the Solar System. Inequality in the Theory of Jupiter and Saturn.

§ **368.** We must now introduce some modification into the facts and the laws we have been asserting. Since all members of the solar system are exposed to one another's influence, and are free to move, they cannot retain the motions which the sun's influence alone would impress on them. Were the planets attracted by the sun only, they would describe perfect ellipses; as each planet and satellite attracts every other planet and satellite, they move in no known or symmetrical curve, but in paths now approaching to, now receding from the elliptical form. Thus we find in the heavens no perfect ellipse, no immovable plane, no unvarying motion; no cubes of the distances bear precisely the same proportion to the squares of the times, no radius vector sweeps over equal areas in exactly equal times. The areas really described, however, the departures from an elliptic path, the alterations of the planes, and the motions of the nodes and the apsides become the tests of the disturbing forces.

An attraction which acts equally and in the same direction on two bodies does not disturb their relative motions. The force which disturbs the motion of a satellite or a planet is the difference of the forces which act on the central and revolving body. Thus if the moon is between the sun and the earth, and if the sun's attraction, in a certain time,

* This chapter is taken from Mrs. Somerville's "Connection of the Physical Sciences."

draws the earth 200 inches, and in the same time draws the moon 201 inches, then the real disturbing force is the force which would produce in the moon a motion of one inch from the earth. If the direction of the attracting force is different in the two cases, some complication is introduced; but by the resolution of forces the amount of disturbance in any given direction may be found.

The disturbing body may be exterior to the orbit of the revolving body, as Jupiter to the earth; or within it, as Venus to the earth; or it may be central and fixed, as the sun, while the two bodies whose relative motions are disturbed both revolve round it.

§ **369.** The simplest mode of considering perturbations is however to consider merely the amount of force and the direction in which it is exerted, without regard to the body exerting it.

To determine the motion of each body, when disturbed by all the rest, is beyond the power of analysis. It is therefore necessary to estimate the disturbing action of one planet at a time, whence the celebrated problem of the three bodies, originally applied to the moon, the earth, and the sun; namely, the masses being given of three bodies projected from three given points, with velocities given, both in quantity and direction; and, supposing the bodies to gravitate to one another with forces that are directly as their masses, and inversely as the squares of their distances, to find the lines described by these bodies, and their positions at any given instant: or in other words, to determine the path of a celestial body when attracted by a second body, and disturbed in its motions round the second body by a third—a problem equally applicable to planets, satellites, and comets.

By this problem the motions of translation of the celestial bodies are determined. It is an extremely difficult one, and would be infinitely more so, if the disturbing action were not very small when compared with the central force; that is, if the action of the planets on one another were not very small when compared with that of the sun. As the disturbing influence of each body may be found separately, it is assumed that the action of the whole system, in disturbing any one planet, is equal to the sum of all the particular disturbances it experiences, on the general mechanical principle, that the sum of any of the small oscillations is nearly equal to their simultaneous and joint effect.

§ **370.** On account of the reciprocal action of matter, the stability of the system depends on the intensity of the primitive momentum of the planets, and the ratio of their masses to that of the sun; for the nature of the conic sections in which the celestial bodies move, depends upon the velocity with which they were first impelled in space. Had that velocity been such as to make the planets move in orbits of unstable equilibrium, their mutual attractions might have changed them into parabolas, or even hyperbolas; so that the earth and planets might, ages ago, have been sweeping far from our sun through the abyss of space. But as the orbits differ very little from circles, the momentum of the planets, when projected, must have been exactly sufficient to insure the permanency and stability of the system. Besides the mass of the sun is vastly greater than that of any planet; and as their inequalities bear the same ratio to the elliptical motions, that their masses do to that of the sun, their mutual disturbances only increase or diminish the eccentricities of their orbits by very minute quantities; consequently the magnitude of the sun's mass is the principal cause of the stability of the system. There is not in the physical world a more splendid example of the adaptation of means to the accomplishment of an end, than is exhibited in the nice adjustment of these forces, at once the cause of the variety and of the order of nature.

§ **371.** The planets are subject to disturbances of two kinds, both resulting from the constant operation of their reciprocal attraction: one kind, depending upon their positions with regard to each other, begins from zero, increases to a maximum, decreases, and becomes zero again, when the planets return to the same relative positions. In consequence of these, the disturbed planet is sometimes drawn away from the sun, sometimes brought nearer to him: sometimes it is accelerated in its motion, sometimes retarded. At one time it is drawn above the plane of its orbit, at another time below it, according to the position of the disturbing body. All such changes, being accomplished in short periods, some in a few months, others in years, or in hundreds of years, are denominated periodic inequalities. The inequalities of the other kind, though occasioned likewise by the disturbing energy of the planets, are entirely independent of their relative positions. They depend upon the relative positions of the orbits alone, whose forms and places in space are altered by very minute quantities, in immense periods of time, and are therefore called secular inequalities.

The periodical disturbances are compensated, when the bodies return to the same relative positions with regard to one another and the sun: the secular inequalities are compensated, when the orbits return to the same positions relatively to one another, and to the plane of the ecliptic.

§ **372.** Planetary motion, including both these kinds of disturbance, may be represented by a body revolving in an ellipse, and making small and transient deviations, now on one side of its path, and now on the other, whilst the ellipse itself is slowly, but perpetually changing both in form and position.

The periodic inequalities are merely transient deviations of a planet from its path, the most remarkable of which only lasts about 918 years; but in consequence of the secular disturbances, the apsides, or extremities of the major axis of all the orbits, have a direct but variable motion in space, excepting those of the orbit of Venus, which are retrograde, and the lines of the nodes move with a variable velocity in a contrary direction. Besides these, the inclination and eccentricity of every orbit are in a state of perpetual but slow change. These effects result from the disturbing influence of all the planets on each. But as it is only necessary to estimate the disturbing influence of one body at a time, what follows may convey some idea of the manner in which one planet disturbs the elliptical motion of another.

§ **373.** Suppose two planets moving in ellipses round the sun; if one of them attracted the other and the sun with equal intensity, and in parallel directions, it would have no effect in disturbing the elliptical motion. The inequality of this attraction is the sole cause of perturbation, and the difference between the disturbing planet's action on the sun and on the disturbed planet constitutes the disturbing force, which consequently varies in intensity and direction with every change in the relative positions of the three bodies. Although both the sun and planet are under the influence of the disturbing force, the motion of the disturbed planet is referred to the centre of the sun as a fixed point, for convenience. The whole force which disturbs a planet, is equivalent to three partial forces. One of these acts on the disturbed planet, in the direction of a tangent to its orbit, and is called the tangential force ; it occasions secular inequalities in the form and position of the orbit in its own plane, and is the sole cause of the periodical perturbations in the planet's longitude. Another acts upon the same body in the direction of its radius vector, that is, in the line joining the centres of the sun and planet, and is called the radial force : it produces periodical changes in the distance of the planet from the sun, and affects the form and position of the orbit in its own plane. The third, which may be called the perpendicular force, acts at right angles to the plane of the orbit, occasions the periodic inequalities in the planet's latitude, and affects the position of the orbit with regard to the plane of the ecliptic.

§ **374.** It has been observed, that the radius vector, of a planet moving in a perfectly elliptical orbit, passes over equal spaces or areas in equal times ; a circumstance which is independent of the law of the force, and would be the same, whether it varied inversely as the square of the distance, or not, provided only that it be directed to the centre of the sun. Hence the tangential force, not being directed to the centre, occasions an unequable description of areas, or what is the same thing, it disturbs the motion of the planet in longitude. The tangential force sometimes accelerates the planet's motion, sometimes retards it, and occasionally has no effect at all. Were the orbits of both planets circular, a complete compensation would take place at each revolution of the two planets, because the arcs in which the accelerations and retardations take place, would be symmetrical on each side of the disturbing force. For it is clear, that if the motion be accelerated through a certain space, and then retarded through as much, the motion at the end of the time will be the same as if no change had taken place. But as the orbits of the planets are ellipses, this symmetry does not hold: for, as the planet moves unequably in its orbit, it is in some positions more directly, and for a longer time, under the influence of the disturbing force than in others. And although multitudes of variations do compensate each other in short periods, there are others, depending on peculiar relations among the periodic times of the planets, which do not compensate each other till after one, or even till after many revolutions of both bodies. A periodical inequality of this kind in the motions of Jupiter and Saturn, has a period of no less than 918 years.

§ **375.** The radial force, or that part of the disturbing force which acts in the direction of the line joining the centres of the sun and disturbed planet, has no effect on the areas, but is the cause of periodical changes of small extent in the distance of the planet from the sun. It has already been shown, that the force producing perfectly elliptical motion varies inversely as the square of the distance, and that a force following any other law, would cause the body to move in a curve of a very different kind. Now, the radial disturbing force varies directly as the distance; and as it sometimes combines with and increases the intensity of the sun's attraction for the disturbed body, and at other times opposes and consequently diminishes it, in both cases it causes the sun's attraction to deviate from the exact law of gravity, and the whole action of this compound central force on the disturbed body, is either greater or less than is requisite for perfectly elliptical motion. When greater, the curvature of the disturbed planet's path on leaving its perihelion, or point nearest the sun, is greater than it would be in the ellipse, which brings the planet to its aphelion, or point farthest from the sun, before it has passed through 180°, as it would do if undisturbed. So that in this case the apsides, or extremities of the major axis, advance in space. When the central force is less than the law of gravity requires, the curvature of the planet's path is less than the curvature of the ellipse. So that the planet, on leaving its perihelion, would pass through more than 180° before arriving at its aphelion, which causes the apsides to recede in space. Cases both of advance and recess occur during a revolution of the two planets; but those in which the apsides advance, preponderate.

§ **376**. This, however, is not the full amount of the motion of the apsides; part arises, also, from the tangential force, which alternately accelerates and retards the velocity of the disturbed planet. An increase in the planet's tangential velocity diminishes the curvature of its orbit, and is equivalent to a decrease of central force. On the contrary, a decrease of the tangential velocity, which increases the curvature of the orbit, is equivalent to an increase of central force. These fluctuations, owing to the tangential force, occasion an alternate recess and advance of the apsides, after the manner already explained. An uncompensated portion of the direct motion arising from this cause, conspires with that already impressed by the radial force, and in some cases, even nearly doubles the direct motion of these points. The motion of the apsides may be represented by supposing a planet to move in an ellipse, while the ellipse itself is slowly revolving about the sun in the same plane. This motion of the major axis, which is direct in all the orbits except that of the planet Venus, is irregular, and so slow that it requires more than 109,830 years, for the major axis of the earth's orbit to accomplish a sidereal revolution, that is, to return to the same stars; and 20,984 years to complete its tropical revolution, or to return to the same equinox. The difference between these two periods arises from a retrograde motion in the equinoctial point, which meets the advancing axis, before it has completed its revolution with regard to the stars. The major axis of Jupiter's orbit requires no less than 200,610 years to perform its sidereal revolution, and 22,748 years to accomplish its tropical revolution, from the disturbing action of Saturn alone.

§ **377.** A variation in the eccentricity of the disturbed planet's orbit, is an immediate consequence of the deviation from elliptical curvature, caused by the action of the disturbing force. When the path of the body, in proceeding from its perihelion to its aphelion, is more curved than it ought to be, from the effect of the disturbing forces, it falls within the elliptical orbit, the eccentricity is diminished, and the orbit becomes nearly circular; when that curvature is less than it ought to be, the path of the planet falls without the elliptical orbit, and the eccentricity is increased; during these changes, the length of the major axis is not altered, the orbit only bulges out or becomes more flat. Thus the variation in the eccentricity arises from the same cause that occasions the motion of the apsides. There is an inseparable connection between these two elements: they vary simultaneously, and have the same period; so that whilst the major axis revolves in an immense period of time, the eccentricity increases and decreases by very small quantities, and at length returns to its original magnitude at each revolution of the apsides. The terrestrial eccentricity is decreasing at the rate of about forty miles annually; and if it were to decrease equably, it would be 39,861 years before the earth's orbit became a circle. The mutual action of Jupiter and Saturn occasions variations in the eccentricities of both orbits, the greatest eccentricities of Jupiter's orbit corresponding to the least of Saturn's. The period in which these vicissitudes are accomplished is 70,414 years, estimating the action of these two planets alone; but if the action of all the planets were estimated, the cycle would extend to millions of years.

§ **378.** That part of the disturbing force is now to be considered which acts perpendicularly to the plane of the orbit, causing periodic perturbations in latitude, secular variations in the inclination of the orbit, and a retrograde movement to its nodes on the true plane of the ecliptic. This force tends to pull the disturbed body above, or push it below the plane of its orbit, according to the relative positions of the two planets with regard to the sun, considered to be fixed. By this action it sometimes makes the plane of the orbit of the disturbed body tend to coincide with the plane of the ecliptic, and sometimes increases its inclination to that plane. In consequence of which, its nodes alternately recede or advance on the ecliptic. When the disturbing planet is in the line of the disturbed planet's nodes, it neither affects these points, the latitude, nor the inclination, because both planets are then in the same plane. When it is at right angles to the line of the nodes, and the orbit symmetrical on each side of the disturbing force, the average motion of these points, after a revolution of the disturbed body, is retrograde, and comparatively rapid; but when the disturbing planet is so situated that the orbit of the disturbed planet is not symmetrical on each side of the disturbing force, which is most frequently the case, every possible variety of action takes place. Consequently, the nodes are perpetually advancing or receding with unequal velocity; but as a compensation is not effected, their motion is, on the whole, retrograde.

§ **379.** With regard to the variations in the inclination, it is clear, that, when the orbit is symmetrical on each side of the disturbing force, all its variations are compensated after a revolution of the disturbed body, and are merely periodical perturbations in the planet's latitude; and no secular change is induced in the inclination of the orbit. When, on the contrary, that orbit is not symmetrical on each side of the disturbing force, although many of the variations in latitude are transient or periodical, still, after a complete revolution of the disturbed body, a portion remains uncompensated, which forms a secular change in the inclination of the orbit to the plane of the ecliptic. It is true, part of this secular change in the inclination is compensated by the revolution of the disturbing body, whose motion has not hitherto been taken into the account, so that perturbation compensates perturbation; but still, a comparatively permanent change is effected in the inclination, which is not compensated till the nodes have accomplished a complete revolution.

§ **380.** The changes in the inclination are extremely minute, compared with the motion of the nodes, and there is the same kind of inseparable connection between their secular changes that there is between the variation of the eccentricity and the motion of the major axis. The nodes and inclinations vary simultaneously, their periods are the same, and very great. The nodes of Jupiter's orbit, from the action of Saturn alone, require 36,261 years to accomplish even a tropical revolution. In what precedes, the influence of only one disturbing body has been considered; but when the action and reaction of the whole system are taken into account, every planet is acted upon, and does itself act, in this manner, on all the others; and the joint effect keeps the inclinations and eccentricities in a state of perpetual variation. It makes the major axes of all the orbits continually revolve, and causes, on an average, a retrograde motion of the nodes of each orbit upon every other. The ecliptic itself is in motion from the mutual action of the earth and planets, so that the whole is a compound phenomenon of great complexity, extending through unknown ages. At the present time, the inclinations of all the orbits are decreasing, but so slowly, that the inclination of Jupiter's orbit is only about six minutes less than it was in Ptolemy's time.

§ **381.** But, in the midst of all these vicissitudes, the length of the major axes and the mean motions of the planets remain permanently independent of secular changes. They are so connected by Kepler's law, of the squares of the periodic times being proportional to the cubes of the mean distances of the planets from the sun, that one cannot vary without affecting the other. And it is proved that any variations which do take place are transient, and depend only on the relative positions of the bodies.

It is true that, according to theory, the radial disturbing force should permanently alter the dimensions of all the orbits, and the periodic times of all the planets, to a certain degree. For example, the masses of all the planets revolving within the orbit of any one, such as Mars, by adding to the interior mass, increase the attracting force of the sun, which, therefore, must contract the dimensions of the orbit of that planet, and diminish its periodic time; whilst the planets exterior to the orbit of Mars must have the contrary effect. But the mass of the whole of the planets and satellites taken together is so small, when compared with that of the sun, that these effects are quite insensible, and could only have been discovered by theory. And, as it is certain that the length of the major axis and mean motions are not permanently changed by any other power whatever, it may be concluded that they are invariable.

§ **382.** With the exception of these two elements, it appears that all the bodies are in motion, and every orbit in a state of perpetual change. Minute as these changes are, they might be supposed to accumulate, in the course of ages, sufficiently to derange the whole order of nature, to alter the relative positions of the planets, to put an end to the vicissitudes of the seasons, and to bring about collisions, which would involve our whole system, now so harmonious, in chaotic confusion. It is natural to inquire what proof exists that nature will be preserved from such a catastrophe? Nothing can be known from observation, since the existence of the human race has occupied comparatively but a point in duration, while these vicissitudes embrace myriads of ages. The proof is simple and conclusive. All the variations of the solar system, secular as well as periodic, are expressed analytically by the sines and cosines of circular arcs, which increase with the time; and, as a sine or cosine can never exceed the radius, but must oscillate between zero and unity, however much the time may increase, it follows that, when the variations have accumulated to a maximum, by slow changes, in however long a time, they decrease, by the same slow degrees, till they arrive at their smallest value, again to begin a new course; thus forever oscillating about a mean value. This circumstance, however, would be insufficient were it not for the small eccentricities of the planetary orbits, their minute inclinations to the plane of the ecliptic, and the revolutions of all the bodies as well planets as satellites in the same direction. These secure the perpetual stability of the solar system.

§ **383.** The equilibrium, however, would be deranged, if the planets moved in a resisting medium sufficiently dense to diminish their tangential velocity, for then both the eccentricities and the major axes of the orbits would vary with the time, so that the stability of the system would be ultimately destroyed. The existence of an etherial fluid is now proved; and, although it is so extremely rare that hitherto its effects on the motions of the planets have been altogether insensible, there can be no doubt, that, in the immensity of time, it will modify the forms of the planetary orbits, and may at last even cause the destruction of our system, which in itself contains no principle of decay, unless a rotary motion from west to east has been given to this fluid by the bodies of the solar system, which have all been revolving about the sun in that direction for unknown ages. This rotation, which seems to be highly probable, may even have been coeval with its creation.

§ **384.** The form and position of the planetary orbits, and the motion of the bodies in the same direction, together with the periodicity of the terms in which the inequalities are expressed, assure us that the variations of the system are confined within very narrow limits, and that although we do not know the extent of the limits, nor the period of that grand cycle, which probably embraces millions of years, yet they never will exceed what is requisite for the stability and harmony of the whole, for the preservation of which every circumstance is so beautifully and wonderfully adapted.

The plane of the ecliptic itself, though assumed to be fixed at a given epoch for the convenience of astronomical computation, is subject to a minute secular variation of 45″ 7; occasioned by the reciprocal action of the planets. But as this is also periodical, and cannot exceed 2° 42′, the terrestrial equator, which is inclined to it at an angle of 23° 27′ 37″ 89, will never coincide with the plane of the ecliptic; so there never can be perpetual spring. The rotation of the earth is uniform; therefore day and night, summer and winter, will continue their vicissitudes, while the system endures, or is undisturbed by foreign causes.

§ **385.** Notwithstanding the permanency of our system, the secular variations in the planetary orbits would have been extremely embarrassing to astronomers when it became necessary to compare observations separated by long periods. The difficulty was in part obviated, and the principle for accomplishing it established by La Place, and has since been extended by M. Poinsot. It appears that there exists an invariable plane, passing through the centre of gravity of the system, about which the whole oscillates within very narrow limits, and that this plane will always remain parallel to itself, whatever changes time may induce in the orbits of the planets, in the plane of the ecliptic, or even in the law of gravitation; provided only that our system remains unconnected with any other. La Place found that the plane in question is inclined to the ecliptic at an angle of nearly 1° 34′ 15″, and that, in passing through the sun, and about midway between the orbits of Jupiter and Saturn, it may be regarded as the equator of the solar system, dividing it into two parts, which balance one another in all their motions. This plane of greatest inertia, by no means peculiar to the solar system, but existing in every system of bodies submitted to their mutual attractions only, always maintains a fixed position, whence the oscillations of the system may be estimated through unlimited time.

§ **386.** Future astronomers will know, from its immutability or variation, whether the sun and his attendants are connected or not with the other systems of the universe. Should there be no link between them, it may be inferred, from the rotation of the sun, that the centre of gravity of the system situate within his mass describes a straight line in this invariable plane or great equator of the solar system, which, unaffected by the changes of time, will maintain its stability through endless ages. But, if the fixed stars, comets, or any unknown and unseen bodies, affect our sun and planets, the nodes of this plane will slowly recede on the plane of that immense orbit which the sun may describe about some most distant centre, in a period which it transcends the power of man to determine. There is every reason to believe that this is the case; for it is more than probable that, remote as the fixed stars are, they in some degree influence our system, and that even the invariability of this plane is relative, only appearing to be fixed to creatures incapable of estimating its minute and slow changes during the small extent of time and space granted to the human race. If we raise our views to the whole extent of the universe, and consider the stars together with the sun, to be wandering bodies, revolving about the common centre of creation, we may then recognise in the equatorial plane passing through the centre of gravity of the universe the only instance of absolute and eternal repose.

§ **387.** All the periodic and secular inequalities deduced from the law of gravitation, are so perfectly confirmed by observation, that analysis has become one of the most certain means of discovering the planetary irregularities, either when they are too small, or too long in their periods to be detected by other methods. Jupiter and Saturn, however, exhibit inequalities, which for a long time seemed discordant with that law. All observations, from those of the Chinese and Arabs down to the present day, prove that for ages the mean motions of Jupiter and Saturn have been affected by a great inequality of a very long period, forming an apparent anomaly in the theory of planets. It was long known, by observation, that five times the mean motion of Saturn is nearly equal to twice that of Jupiter; a relation which the sagacity of La Place perceived to be the cause of a periodic irregularity in the mean motion of each of these planets, which completes its period in 918 years, the one being retarded while the other is accelerated; but both the magnitude and period of these quantities vary, in consequence of the secular variations in the elements of the orbits. Suppose the two planets to be on the same side of the sun, and all three in the same straight line, they are then said to be in conjunction. Now if they begin to move at the same time, one making exactly five revolutions in its orbit, while the other only accomplishes two, it is clear that Saturn, the slow moving body, will only have got through a part of its orbit during the time that Jupiter has made one whole revolution and part of another, before they be again in conjunction.

§ **388.** It is found that during this time their mutual action is such as to produce a great many perturbations which compensate each other, but there still remains a portion outstanding, owing to the length of time during which the forces act in the same manner ; and if the conjunction always happened in the same point of the orbit, this uncompensated inequality in the mean motion would go on increasing till the periodic times and forms of the orbits were completely and permanently changed ; a case that would actually take place if Jupiter accomplished exactly five revolutions in the time Saturn performed two. These revolutions are, however, not exactly commensurable ; the points in which the conjunctions take place are in advance each time as much as 8°. 37 ; so that the conjunctions do not happen exactly in the same points of the orbits till after a period of 850 years ; and in consequence of this small advance, the planets are brought into such relative positions that the inequality which seemed to threaten the stability of the system is completely compensated, and the bodies having returned to the same relative positions with regard to one another and the sun, begin a new course. The secular variations in the elements of the orbit increase the period of the inequality to 918 years. As any perturbation which affects the mean motion affects also the major axis, the disturbing forces tend to diminish the major axis of Jupiter's orbit, and increase that of Saturn's during one half of the period, and the contrary during the other half. This inequality is strictly periodical, since it depends on the configuration of the two planets ; and theory is confirmed by observation, which shows that, in the course of twenty centuries, Jupiter's mean motion has been accelerated by about 3° 23′, and Saturn's retarded by 5° 13′.

§ **389.** Several instances of perturbations of this kind occur in the solar system. One, in the mean motions of the Earth and Venus only amounting to a few seconds, has been recently worked out with immense labor by Professor Airy. It accomplishes its changes in 240 years, and arises from the circumstance of thirteen times the periodic time of Venus being nearly equal to eight times that of the earth. Small as it is, it is sensible in the motions of the earth.

It might be imagined that the reciprocal action of such planets as have satellites would be different from the influence of those that have none. But the distances of the satellites from their primaries are incomparably less than the distances of the planets from the sun, and from one another. So that the system of a planet and its satellites, moves nearly as if all these bodies were united in their common centre of gravity. The action of the sun, however, in some degree disturbs the motion of the satellites about their primary.

CHAPTER XVII.

PRECESSION, NUTATION, AND ABERRATION.

Action of the Planets on the Plane of the Ecliptic. Action of the Sun and Moon on the Earth's Equator. The Precession of the Equinoxes. Motion of the Earth's Axis. Nutation. Aberration. Its Effect on the Apparent Places of the Stars. Methods of Computing it. Aberration of the Fixed Stars and of the Planets.

§ **390.** The planets acting on the sun and on the earth, as a whole, occasion a slow variation in the plane of the ecliptic, which affects its inclination to the plane of the equator, and but for other causes would make the equator cross the ecliptic every year 0.″31 in advance of the last equinox. The disturbing influence of Venus and Jupiter in particular on the earth diminishes the obliquity of the ecliptic annually by 0″457. This variation, in the course of ages, may amount to 10° or 11°; but the obliquity of the ecliptic to the equator can never vary more than 2° or 3°, since the equator will follow in some degree the motion of the ecliptic.

But while the action of the planets would cause the equator to cross the ecliptic later every year, more powerful influences are drawing them to intersect earlier. The earth, it will be remembered, is not a perfect sphere. The sun and moon being always, one in, and the other near, the plane of the ecliptic, act obliquely and unequally on different parts of the spheroid, and urge the plane of the equator backward from east to west. There are but two positions (in the equinoxes) when the sun does not urge the earth's equator to change its position. At all other times, by its action on the protuberant matter at the equator, it tends to draw the equatorial parts under itself, and thus causes a balancing of the equator.

§ **391.** The direct tendency of this attraction is to make the planes of the equator and ecliptic coincide. It is difficult to imagine or compute the effects of this tendency, for the rapid rotation of the earth continually presents different parts to the sun, and thus changes the direction, and consequently the force of the sun's attraction on the equatorial parts. And the equatorial parts cannot obey the attraction without drawing along the whole mass of the earth. Therefore, the motion of the whole earth, which can be caused by the excess of attraction on the equatorial parts at one moment, is extremely small.

The inclination of the planes is not affected by this cause; but in the course of the year sufficient attraction is exercised by the sun to cause the equator to cut the ecliptic in the vernal equinox, at a point 15″ sooner than it would if the earth were a perfect sphere. This recession of the equator causes the precession of the equinoxes; that is, causes the equinox to occur sooner than it would if the earth were a perfect sphere. If the flattening of the earth were greater than it is, precession would also increase.

The orbit of the moon being inclined 5°.8 to the ecliptic, is sometimes inclined 29° to the earth's equator. The moon's action on the redundant matter at the equator has a similar effect to that of the sun.

§ **392.** The actual yearly recession caused by the sun and moon would be rather more than 50″ 22, but it is partly balanced by the action of the planets mentioned before. Thus the sun, moon and planets, by moving the plane of the equator, cause the equinoctial points to retrograde on the ecliptic, without however in the whole changing the angle made by the equator and ecliptic. And the planets move the plane of the ecliptic, and give the equinoctial points a direct motion, though much less than the former.

The direct action of the planets alone on the equatorial parts of the earth causes a slight retrogradation of 50″ a century.

I have mentioned that even while the planets act on the earth independently of its figure, they have an indirect influence connected with this figure. By displacing the plane of the ecliptic, they bring the sun and moon into different positions with respect to the earth, and thus modify their action on the equatorial parts. These inequalities render the retrogradation unequal in different centuries.

The period in which the equinoctial points would accomplish an entire revolution in the ecliptic, cannot therefore be precisely fixed. It does not vary much from 25,868 years.

§ **393.** If the equator retreats on the ecliptic, the pole of the equator must move also, and describe round the pole of the ecliptic from east to west, a small circle with a radius of 23° 28′, in 25,868 years.

If the attraction of the sun and moon were always exercised equally, this would be the path described by the pole of the equator. But the inclination of the moon's orbit to the earth's equator, and the position of her nodes are continually changing. She is also rapidly changing place in her orbit, and her orbit is changing its line of apsides, and consequently its point of nearest approach to the earth. The position of the nodes, however, has more influence than the place of the moon in her orbit. For, by this position, we ascertain in which direction the moon's attraction on the equatorial parts will preponderate in the course of a lunar month. To this balancing motion of the equator, and the wavering motion of the pole, caused by the inequalities of the moon's action, the name of nutation is given. Were the pole of the earth influenced by the disturbing action of the moon only, it would describe about the pole of the moon's orbit a small ellipse, the axes of which are 18″ 5, and 13″ 6, the longer being directed toward the pole of the ecliptic. If this were described alone its period would be 19 years, the time occupied by the nodes of the lunar orbit in accomplishing a revolution.

§ **394.** To give an idea of the extreme minuteness of

the nutation of the earth's axis caused by the moon, let us suppose an iron rod 100 feet long, fixed at one end and movable at the other, to represent one half of the earth's axis. If the movable end were pulled the twentieth part of an inch to one side, the deviation would be proportionally as great as that which the lunar nutation produces in the terrestrial axis. There is also a slight inequality which depends on the position of the sun only, which goes through all its values in the course of half a revolution of the sun. On account of it the pole would describe an ellipse, whose semi-major axis would be 0″ 435, its semi-minor axis 0″ 399.

The curve really traced in the heavens by the prolongation of the earth's axis is compounded of these three motions. While in virtue of the moon's action it would describe a little ellipse, it is carried over so much of its circle round the pole of the ecliptic as corresponds to 19 years ; that is to say, over an angle of nineteen times 50″ round the centre. The path which it will describe in virtue of these three motions will be neither an ellipse nor an exact circle, but a slightly undulating ring. In the following figure the ellipse caused by the sun's action is not represented.

§ **395**. Let C, (Plate II. Fig. 4), be the centre of the earth, C P half its axis, P the north pole, and A S B half of the equator. Let m n be part of the plane of the ecliptic, and C Q a line perpendicular to it, pointing therefore to the pole of the ecliptic in the heavens. If then P be carried uniformly round a circle perpendicular to C Q, so that C P shall describe a conical surface, the equinoxes B and A will be carried round in a direction contrary to that of the diurnal motion, and with them the equator B S A, the angle which the equator makes with the ecliptic remaining unaltered. This motion of B and A is the precession. But suppose that instead of P's being placed on the circle, it is placed on the circumference of a small oval which has its centre on the circle. While the centre of the oval moves forward on the circle with the motion of precession, let the pole P move round the oval with a motion much slower than that of precession. It will then trace out in space an undulating curve, and there will be an alternate retardation and acceleration of the motion of the equinoxes along the plane of the ecliptic, together with a vibration of the plane of the equator to and from the ecliptic, which are the motions constituting nutation.

§ **396.** We must now make a slight change in our conception of the motion of the earth, and imagine its centre moving on in the ecliptic, while at the same time all the parts but the axis of the equator rotate. This axis is not wholly without motion. Its middle point, to be sure, is always stationary, but the ends of the axis describe two circles blended, one owing to precession, one owing to nutation; the cause of these motions we have just seen; let us now view them separately, without considering the especial share which the sun, moon, and planets have in each.

If our eyes were keen enough to discern exceedingly slow motion, or if we could see at one glance the motion of years, we should see the equinoxes moving back on the ecliptic, with a variable motion. At the same time the equator would appear to swing backward and forward to and from the ecliptic, turning upon the equinoxes as pivots. Of these motions the average motion of the equator on the ecliptic is the precession; the alternate acceleration and retardation are one part of the nutation; and the alternate increase and diminution of the angle contained between the two, is the other part.

§ **397.** As the fixed stars do not alter their positions among themselves in consequence of the retrocession of the ecliptic, by observing when the equator passes through them, it is easy to ascertain the amount of its motion. This being known, we know the position of the vernal equinox at that time, a point which it is essential to know, since we reckon from it longitudes and right ascensions, and also begin there our equinoctial year. When we speak of the right ascension or declination of any celestial object, we must therefore mention what epoch we intend.

Precession affects the longitudes of all the stars, their declinations and right ascensions, but not their latitudes.

The right ascensions and the declinations of different stars are variously affected by the motion of the pole of the equator. Some stars and constellations are brought by precession near to the pole, and others appear to recede from it. Its effects on the places of the stars are so striking, that it was discovered more than a century before the Christian era, though its cause remained unknown. Nutation, which makes but inconsiderable changes among the stars, has been known but little more than a hundred years. Nutation causes an apparent approach and recess of all the stars in the heavens to the pole in the period of nineteen years. The equinoctial points have also a small alternate balancing motion in the same period, by which the longitudes and right ascensions of the stars are alternately increased and diminished.

§ **398.** In the year 158 before Christ, Hipparchus discovered precession from the comparison of his own observations with those made 155 years before. He had formed a catalogue in which he laid down the latitude and longitude of every star; and he supposed that this work once performed would give the true places of the stars forever. In his own lifetime, however, he found that all the fixed stars were sweeping with a very slow motion towards the east, and that while their latitudes remained the same, their longitudes would increase the 360th part of the whole circumference in 72 years. He had no hesitation between attributing a real motion to the stars or one in the opposite direction to the earth.

§ **399.** It is shown, both by observation and theory, that the orbits of all the planets, as well as that of the earth, have this motion by which their nodes move on the ecliptic, and on the imaginary plane of inertia before mentioned. They do not, however, all retreat on the plane of inertia. Those whose orbits are most inclined to this plane are drawn down by those whose orbits are less inclined, and their nodes advance; but those whose orbits are more nearly in this plane, are drawn from it by those which are farther removed, and their nodes consequently retreat. Even those orbits which actually retreat on the plane of inertia, may appear to advance on that part of any orbit which lies above this plane.

In all the planets, except Venus, there is a very little more than a complete revolution between two aphelia; in Venus, there is a little less. The apparent annual motion of the aphelion of each planet's orbit is the motion arising from precession plus that arising from the motion of the apsides. In Venus, the motion arises from precession minus the motion of the apsides. The apparent motion of Venus's aphelion is like that of the other planets; for though the aphelion moves backward, the equinox does the same at a greater rate. The real motion of its apsides is regressive, because its orbit is very nearly circular, and the amount of the earth's influence is to make it regress.

§ **400.** Precession and nutation afford us some light as to the condition and density of the interior of the earth. The mean density of the earth being little more than 5½ times the density of water, and the rocks on the surface averaging only about 2½ times the weight of water, it follows that the interior must be as much above the average as the surface is below it.

The ring of matter round the equator being composed partly of rock and partly of water, cannot average above 2½ times the weight of water. Its effect in disturbing the earth's axis, is therefore slighter than if it were of the same density with the interior. The disturbance found by calculation is less than if the earth were a homogenious spheroid, we are therefore justified in believing it not to be one. We find, also, that the comparative densities of the interior and exterior, found by the disturbance of a pendulum near a mountain, agree with those inferred from nutation.

The observed amount of precession seems to require for the crust of the globe a greater thickness than many geologists have allowed. It requires a thickness of at least one fourth or one fifth of the earth's radius; that is, from eight hundred to a thousand miles. It does not forbid the supposition that the earth is solid to the centre.

§ **401.** There is another phenomenon besides the change in longitude of the stars, which early made the precession of the equinoxes suspected. The sun was found to cross the equinox before he returned to the same stars. In other words, the equinoctial was found to be shorter than the sidereal year. If there were no precession, the equinoctial and sidereal year would agree. As it exists, the equinoctial year of 365d. 5h. 48m. 49s. must be increased by the time the sun takes to move through an arc of 50″22, in order to have the length of the sidereal year. The time required is 20′ 19″; so that the sidereal year contains 365d. 6h. 9m. 9s., mean solar days.

Owing to the variations in the action of the sun, and the consequent inequalities in the precession of the equinoxes, the equinoctial year is four or five seconds shorter than it was in the time of Hipparchus. The annual retrogradation of the equinoctial points being greater by 0.″455 than it was in his time, the sun has each year a space of 0.″455 less in the ecliptic to pass through, in order to reach the plane of the equator. The utmost change in the length of the year from this cause is 43 seconds.

§ **402.** Besides the effect of precession and nutation, another cause influences the apparent positions of the stars. This is the aberration of light.

We judge of the position of bodies by the direction of rays which enter our eyes, and which appear to proceed from them. We know not how often these rays have been reflected or refracted, whether they have been left behind by some body rushing on in its course, or whether some motion of our own has caused them to meet our eye in a false direction. A ray from a lantern suspended above our heads may reach us at the same moment and from the same direction with a ray from a star, and the two objects will be referred to the same place. The lantern partakes our motion, and is seen in its true place. But the rays from the star are, by our motion, referred to a false direction.

§ **403.** Even when we are conscious of our motion we are apt to transfer it to surrounding objects. A man walking fast in a vertical rain, attributes to the rain his own motion. If he goes toward the north, the rain appears to him to slant toward the south. If he moves as fast as the rain falls, he gives to the rain as much motion southward as it has vertically, and it strikes him as if it fell at an angle of 45°. A train of cars moving rapidly against a driving rain, gives its own motion to the drops, and they appear to the passengers more slanting than they really are. If the cars move with the rain, they meet the drops sooner than if at rest, and their apparent obliquity is less. Whatever may be the direction in which drops would fall upon cars at rest, they will meet the cars in motion in a different one. By the same reasoning, we see that whatever may be the direction in which rays would pass from a star to an observer at rest, they must meet an observer in motion in a different one.

It remains only to ascertain the amount of this displacement. It depends on the relative velocities of light and of the earth. If light moved with infinite speed, there would be no aberration. Rays would leave the star and reach our eyes before we had changed our place. But its velocity bears a definite ratio to the velocity of the earth in her orbit, 192000 : 19. To ascertain the angle of aberration we should then construct a triangle, of which one side (the velocity of the ray) should be to the other (the velocity of the earth), as 10105 : 1. The angle made by the ray and orbit being also known, we can find the other angles, and the direction of the third side of the triangle.

§ **404.** Let S (Fig. 5, Plate II.) represent a star. Let S B be the actual course of a ray. And, for simplicity, let S B be perpendicular to A B, the line of the earth's motion. The side A C represents the inclination which must be given to the eye, in order that when it has passed on to B, it may meet the ray from S.

B C : B A : : vel. light : vel. earth : : rad. : tang. 20.″246.

Thus the angle B C A, or its equal S C E, by which the direction of the observer's eye deviates from the true direction of the star=20.″246. Also C B D, the displacement of the star, or the amount of aberration=20.″246.

Passing from A to B, the observer has gained upon the ray an angle A C B. Unconscious of his motion, he adds this angle to the inclination of the ray, as the man in the rain adds the angle made by his own motion to the true angle of the rain.

The same reasoning holds good when the direction of the earth's motion is not perpendicular to the ray. The star may be so placed that S B A shall be an acute angle, or so that it shall be an obtuse angle. In either case we have the proportion:

B C : B A : : sine of B A C : sine of A C B, or C B D, (the apparent displacement.)

Thus it appears that the sine of the aberration is proportional to the sine of the angle made by the earth's motion in space with the ray, and is therefore greatest when the line of sight is perpendicular to the orbit.

§ **405.** Aberration distorts the aspect of the heavens, causing all the stars to crowd as it were toward that point in the heavens which is the vanishing point of all lines parallel to that in which the earth is for the moment moving. As the earth moves round the sun in the plane of the ecliptic, this point lies in that plane, 90° in advance of the earth's longitude, or 90° behind the sun's, and moves onward continually, describing the circumference of the ecliptic in a year. It causes each particular star apparently to describe a small ellipse in the heavens, having for its centre the point in which the star would be seen if the earth were at rest.

Let us consider the manner in which the phenomena of aberration succeed each other in the course of the year. Let A B a C, (Fig. 6, Plate II.), represent the celestial ecliptic, S the position of any star, and A S a the half of a great circle drawn through the star perpendicular to the plane of the ecliptic: and let B, C, be other points in the celestial ecliptic, and B S b, C S c, arcs of great circles drawn through the star and these points respectively, which will of course be semicircles, as all great circles bisect each other. If then A, B, C, a, b, c, represent different positions of the point toward which the earth moves, (or the point 90° before the earth's place), S A, S B, S C, S a, S b, S c, will represent the arcs, to the sines of which the amount of aberration is proportional, and in the direction of which it takes place.

§ **406.** Now the earth being in every point of the ecliptic in the course of the year, every point of the celestial ecliptic must be 90° before its place in the course of the same period, and consequently the point S must be joined with every point of the celestial ecliptic to give all the arcs which determine the magnitude and direction of the aberration during the year. Of these the least is A S, the greatest S a; the one being greater and the other less than 90°: and, in passing from A towards a, the corresponding arcs must pass through all intermediate values, increasing as they approach a. Of course, among these, there must be one which is of 90°; let S C be this; and the aberration in the direction of the line S C will have its greatest value, and will of course be 20.″246. The arcs C s, S c, together make a semicircle, and S C being 90°, S c will be 90° also; of course, therefore, the aberration in the direction S c will also be 20.″246; and the extreme distance between the two apparent places as affected by the aberration in these opposite directions will be 40.″492. If again, B S b represent any other great circle, passing through S, the aberration in the direction S B will be proportional to the sine of S B, and that in the direction of S b will be proportional to the sine of S b. But the arcs S B, S b, together make up a semicircle, or S B is the supplement of S b: and as the sine of the arc and of its supplement are equal, the abberations in the directions S B, S b, are equal. In the same manner, the aberrations in the directions S A, S a, are equal; and as S A is the least possible arc drawn from S to the celestial ecliptic, these are the least values of the aberration. The greatest values of the aberration are in the directions S C, S c; and as the arcs S C, S c, are of 90° each, the circle C S c cuts the circle A S a at right angles; or the directions of the greatest and least aberrations are perpendicular to each other, the least aberration taking place in a direction perpendicular to the ecliptic, or affecting only the latitude of the star. On investigation of the precise amount of the aberration in every direction, on the supposition that it is the effect of the earth's motion, we shall find that the apparent place of the star is always in the periphery of the ellipse of which the centre is the true place of the star; the minor axis is in the direction of a great circle passing through the star perpendicular to the ecliptic, the major axis=40.″492; and the proportion of the minor to the major axis, that of sine of star's latitude to radius. Of course, therefore, the star is never seen in its true place, except in one case, which we shall presently mention.

§ **407.** It was said that aberration of light caused stars to move in an ellipse. This is true of all stars except those which are in the ecliptic, or of one in the pole of the ecliptic. Let us suppose the case of a star placed exactly in the pole of the ecliptic. In this case, the arc drawn from the star to every point in the ecliptic is exactly 90°, and its sine, therefore, always equal to the radius. Consequently the star will be seen in a curve parallel to the ecliptic; it will always appear 20.″246 distant from its true place, and the amount of aberration will be always the same, and always the greatest possible. The star will always be 90° further advanced in its orbit than the earth in its orbit. In the case of a star situated in the ecliptic, an arc drawn from the star toward the point of the ecliptic towards which the earth is moving, will always be a portion of the ecliptic itself. The whole aberration therefore will be in the plane of the ecliptic, or will take place entirely in longitude. The magnitude of this arc also will have every value from 0° to 180°; therefore the aberration will, at two points where the arc is 90°, have its greatest value of 20.″246. And at other two, where the arc is 0° or 180°, that is to say, when the earth is moving directly towards or away from the star, the aberration will be nothing. The star's latitude being nothing, the minor axis of the ellipse in which it is seen becomes nothing also, and the ellipse itself becomes a straight line, in which the star appears to oscillate backwards and forwards; passing through the centre, or having its apparent coincide with its true place in the course of its passage each way. All stars which are neither in the ecliptic nor the poles of the ecliptic, describe ellipses of various forms, as described above.

§ **408.** The aberration we have now been considering, arises from the motion of the observer. We will now consider that which arises from the motion of the body observed; its consequences are much more easily conceived of, and it affects only a few of the binary and multiple stars whose motions have been watched. To distinguish the two kinds of aberration, the former has been called subjective, the latter objective aberration. Objective aberration then applies to those apparent displacements which originate in the length of time occupied in the transmission of light from the luminary to the eye.

§ **409.** Let us consider a Binary Star, consisting of two stars, A and B, at so great a distance that light requires x years to reach the earth. Suppose, for simplicity, the common proper motion of the centre of gravity to be in a direction perpendicular to the visual ray. It is obvious that each of the two stars A and B will be seen, independent of the other at any given moment, not in the place which it occupies at that moment, but in that which it did occupy x years since, without regard to any change which may have taken place in its velocity or direction. Since this is true of each individual, it is true of both together, regarded as forming a compound luminary, the parts of which must have, with respect to each other and to the spectator, an apparent situation identical with their real situation x years ago. We see, therefore, the compound object A and B in the state in which it really did exist x years previously. This being true at every instant, it follows that in viewing such a system continually for a series of years, we necessarily perceive its orbit in its true form; and all the angles of position and distances in that orbit will be given truly by our measurements, unaffected by any optical illusion or distortion whatever, only for an epoch antecedent by x years.

§ **410.** There is a curious difference between the consequences of these two kinds of aberration, as shown in the heavens. Subjective aberration prevents our ever seeing a star at any moment in its proper place. We see it displaced by turns on every side. Objective aberration never removes a star from its true course, it merely causes it to lag behind and appear in a place historically, but not at the moment, true.

The sun, moon, and planets suffer both subjective and

objective aberration. Their subjective aberration may be found in the same way as that of the stars. Their objective aberration differs from that of the stars in this, that as we know their distances, we may ascertain how much they appear to lag behind; we may know how long since the place they now appear in was the true one.

The sun being always in the celestial ecliptic, and at the distance of 180° from the earth, he is always 90° before the point toward which the earth is moving; the aberration, therefore, in his case has always its greatest value, of 20.″246. It always affects the longitude only, being in the plane of the ecliptic; and it always diminishes the apparent longitude, because it brings the sun nearer the point toward which the earth is moving. We always, therefore, by the effect of aberration, imagine the sun to be in a point 20.″246 behind the true direction of the ray by which we see him.

The apparent places of the planets are affected in an analogous manner. The true place of any planet at the time of observation differs from the observed place by the arc which the planet describes in the time that a ray of light takes to pass from it to the earth. We must also allow for the motion of the earth during the same time.

§ **411.** In these calculations we have considered the earth's motion in its orbit only, and have placed the observer in the line joining the earth's centre and the sun. There must however be some aberration produced by the earth's rotation. The greatest possible velocity from this cause is that of a point in the equator, .2916 of a mile a second. The greatest aberration from this cause will be to that arising from revolution in proportion to their respective velocities: 19 : .2916 : : 20.″246 : .3108 of a second, a quantity too small to be regarded.

We have also supposed the earth's motion in her orbit to be circular and uniform. The variations from these causes do not exceed 0.″003. It would therefore be an unnecessary refinement to allow for them.

CHAPTER VIII.

TIME.

Natural Divisions of Time. The Solar and Sidereal Day. Mean and Apparent Time. The Equation of Time. Variation in the Length of the Seasons. The Sidereal, Equinoctial and Anomalistic Years. Leap Year. Further Divisions of Time.

§ **412.** By seeing things change, and have a beginning and end, we acquire the idea of before and after. By equal intervals of time we mean successions of events during which we can execute the same things in the same manner, or during which the same phenomena are reproduced in the same order. Unequal intervals are those in which we cannot do the same things, traverse the same road, or perform the same labor. It is then motion which gives the idea of time, and by motion time is measured; for we cannot trust our senses to measure either. Time and motion are the only measures of each other. We can describe a time only by mentioning what portion of her daily or yearly course the earth has performed while it lasts; we can describe the swiftness of a motion only by saying how much time it occupied. Thus the motion which we describe as occupying a given time must ultimately be referred to the time the earth occupies in another motion.

§ **413.** As we have described the motion of a body by saying that it moves a certain distance in a certain time, so we may now divide such portions of time as we can grasp into definite periods during which certain motions last. We cannot conceive of the beginning or end of time, but we can form a definite idea of periods of time. The most obvious periods are those of the revolution and the rotation of the earth, a year, and a day. But we find more than one kind of day, and more than one kind of year. A day, in common language, is the period between two successive appearances of the sun on the meridian, or between his rising and setting. A year is the period during which all the seasons return ; it may be reckoned from mid-winter to mid-winter, or from the vernal equinox to the vernal equinox, or from a point at a certain distance from those points to the same again. But these divisions which are so obvious, and which regulate the labors of the husbandman and the common operations of life, are not nearly precise enough for astronomical purposes. The returns of the heavenly bodies to the meridian divide the year into convenient portions, the return of the sun to the same place in the heavens serves to mark when a larger portion of time has elapsed, but we must have some invariable standard of time, independent of the motions of the earth, with which to compare them. Such a standard we find in the oscillations of the pendulum. For the oscillations of a pendulum of a certain length, in a given latitude, must, in consequence of gravity, always occupy equal portions of time. Having this standard we may now ascertain whether these natural periods are always of the same length.

§ **414.** The two natural days are the solar and the sidereal. Of these the solar day is most convenient for common use, because every one knows how often in the year, and when, the sun is on the meridian, and his presence there, and his rising and setting, control all our movements. But the sidereal day has this advantage, it is invariable.

As the orbit of the earth is but a point compared to the fixed stars, the revolution of the earth does not change the length of a sidereal day. A place on the earth which is under a certain star at noon one day, is again under it in 23h. 56m. 4″ ; and returns to it after the same interval throughout the year. The different portions of each day are also described in proportional periods, as is shown by the apparent motion of the stars. If two stars are in the same circle of rotation, but one is 180° distant from the other, half of a sidereal day elapses between their appulses to the meridian. If they are 90°, a quarter of a sidereal day elapses; and so on for every proportion of distance. Therefore not only is the duration of a sidereal day constant, but during every part of it rotation goes on uniformly. The pendulum also confirms these results.

§ **415.** A solar day is longer than a sidereal day. For while the earth has been once rotating, she has advanced in her orbit nearly one degree, and this makes the sun appear to have advanced one degree. A place therefore which has a certain star and the sun on its meridian one day at noon, must turn round further to bring the sun again on its meridian than to bring the star. Thus an absolute turn of the earth on its axis falls short of a natural day, and the earth requires as much more than one turn on its axis as it has gone forward in that time, which on an average is $\frac{1}{365}$ part of a circle. Hence in 365 days the earth turns 366 times on its axis. Thus there is one more sidereal day in the year than there are solar days of the earth or of any other planet; one turn being lost by each planet's motion round the sun. From a similar cause the traveller who journeys eastward round the world loses one day let him take what time he will.

§ **416.** We find by the pendulum that the solar days are not equal in length one to another, nor are they at all seasons of the year uniformly described. Yet the sidereal day will not answer to regulate the employments of life. For as the sun moving continually eastward is in the course of the year at all distances east of the star chosen to mark the noon of a sidereal day, if we reckoned by the star we should in the course of the year have noon when the sun was rising or setting, or at midnight. The sidereal day is therefore useless for common life, although it is employed by astronomers.

We can, however, take the average of the solar days throughout the year, and thus obtain a mean solar day whose commencement never differs from that of the actual solar day much more than 16 minutes. The mean solar day is divided into 24 hours, each hour into 60 minutes, each minute into 60 seconds, and these are each of a fixed and determinate length. A pendulum 39.13929 inches, in the latitude of London, 51° 31′ 1″, in a vacuum at the level of the sea, vibrates seconds. The pendulum of an astronomical clock is usually made of such a length as to vibrate sidereal seconds. For the sidereal day may likewise be divided into hours, minutes, and seconds. Computations and observations made by sidereal time are easily transferred to solar time, and the reverse.

§ **417.** The common day begins at midnight, but the astronomical day begins at noon, and is counted from 0 to 24 hours. When it is noon, however, in one part of the world, it must be a different hour in all others not under the same meridian. Hence arises great inconvenience, particularly as regards places situated widely apart in longitude; the observed time must be referred to some common epoch. Time is therefore by astronomers referred to a fixed instant common to all the world, to the moment when the mean sun enters the mean vernal equinox. It is reckoned in mean solar days and parts of a day. Time thus reckoned is called equatorial time, and is numerically the same at the same instant in all parts of the globe.

Sidereal time is calculated from the moment when the vernal equinox is on the meridian, and is also counted from 0 to 24 hours. Clocks showing sidereal time are so regulated that they show 0h. 0′ 0″, the instant the equinoctial point passes the meridian of the observatory. And as time is a measure of angular motion, the clock gives the distances of the heavenly bodies from the equinox by observing the instant at which each passes the meridian, and converting the interval into arcs at the rate of 15° an hour.

§ **418.** The variation in length of the solar day arises from two causes; from the equator's not coinciding with the ecliptic, and from the unequal motion of the earth in its orbit. We will consider each of these causes independently of the other. The earth's motion on its axis being perfectly equable, and the plane of the equator being perpendicular to its axis, it is evident that in equal times equal portions of the equinoctial pass the meridian; and so might equal portions of the ecliptic, if the ecliptic were parallel to, or coincident with, the equinoctial. But as the ecliptic is oblique to the equinoctial, the equable motion of the earth carries unequal portions of the ecliptic over the meridian in equal times; near the solstices smaller portions of the ecliptic rise in equal times than near the equinoxes.

If on an artificial globe small patches are placed at every 15° of the equator and of the ecliptic, turning the globe slowly westward, all the patches from Aries to Cancer, come to the brass meridian sooner than the corresponding patches on the equator. All those from Cancer to Libra will come later to the meridian than their corresponding patches on the equator; those from Libra to Capricorn sooner, and those from Capricorn to Aries later. And the patches at the beginning of Aries, Cancer, Capricorn, and Libra, will be either on or even with those on the equator. Thus from Aries to Cancer, the apparent time, as influenced by this cause only, will anticipate the mean time; from Cancer to Libra mean time will be the fastest; from Libra to Capricorn apparent time will again be faster, and from Capricorn to Aries it will be slower than mean time. If the ecliptic were more oblique to the equator than it is there would be still more variation.

§ **419.** The reduction of the apparent place of the sun at any time to its mean place is called the equation of time. Astronomical observations are presented not as they were observed but as they would have been observed if periodical causes of fluctuation had not existed. Getting rid of these fluctuations is termed equating or correcting the observation; the amount added to or subtracted from the mean time is called the equation.

§ **420.** We will now consider the second cause of the inequality in the length of solar days; this is the varying rapidity of the earth in different parts of her orbit. Near the aphelion her daily arc is no more that 57′ 12″, near the perihelion it is 1° 1′ 19″. From Aries to Libra her motion is more rapid than from Libra to Aries. At the aphelion apparent and equated time agree; but the earth advancing from these with less than her mean speed, a place on her surface is brought under the sun in less than twenty-four hours, and apparent time is in advance of equated time shown by the clock. These differences of time increase until the increased speed of the earth begins to diminish them, and finally, at the perihelion, real overtakes apparent time. After passing the perihelion, the earth, moving with her utmost speed, gains upon the sun, and must make more than one rotation to bring the sun on the meridian of a given place. The difference between apparent and equated time increases a while, until diminished by the earth's slackened speed, and at the aphelion the times again coincide. Thus, owing to the excentricity of the earth's orbit, apparent time is in advance of mean time from mid-summer to mid-winter, and behind it from mid-winter to mid-summer.

If the earth moved equably, but in the ecliptic, there would be no equation of time at the equinoxes and the solstices. If it moved with its elliptic irregularity, but in the equinoctial instead of the ecliptic, there would be no equation of time at the aphelion and perihelion. Owing to these combined causes time presents a very irregular series of phenomena in the course of the year. The equation vanishes only when the effect of one cause of irregularity is equal and opposite to the other. This takes place only four times a year; on or near April 15th, June 15th, September 1st, and December 24th. As the perigee is in advance of the solstice, the two causes act partially together in the autumn, while in the spring they partially counteract each other.

Since the hours of daylight extend equally on each side of the apparent noon of each day, the interval from sunrise to twelve o'clock must sometimes be longer and sometimes shorter than that from twelve o'clock to sunset. After the winter solstice of the northern hemisphere, when apparent time lags daily more and more behind the clock, and the hours of daylight are increasing, the sun rises at the same hour many days in succession, but sets later and later. Before the solstice, apparent time being in advance, and the differences between it and real time decreasing daily, the days appear to shorten in the morning only.

§ **421.** The results obtained above depend entirely on the relative positions of the aphelion and the equinoxes. If these are fixed points, or always retain the same relative position, those results will serve alike for every year. If they vary, the equation of time will vary also. Any variation in the inclination of the ecliptic to the equator would also affect it. In fact the inclination of the ecliptic to the equator does undergo some slight variation, not important in its results, but furnishing one reason why calculations for the equation of time do not apply accurately except to the particular periods for which they were computed. The question whether the equinoxes and the aphelion are fixed points is more important, particularly in connection with the division of time into years.

In seeking the relative position of the equinoxes and the perihelion we must remember that the equinoctial point is not stationary but slowly receding from east to west. The aphelion has also a slow motion of 11″66 eastward. The aphelion one year lies 11″66 east of the aphelion of the year before. And as this motion is in an opposite direction from that of the equinox, the difference between the equinox and the aphelion is constantly increasing by the amount of the two motions, or 61″9, annually. At this rate it would accomplish a revolution in the course of 20,984 years. The major axis of the solar ellipse must have coincided with the line of equinoxes about 4,000 years before the Christian era, much about the time which chronologists assign for the creation of man. In 6433 its major axis will again coincide with the line of the equinoxes, but then the solar perigee will coincide with the autumnal equinox, whereas at the creation of man it coincided with the vernal equinox.

§ **422.** The variation in the position of the solar ellipse occasions corresponding changes in the length of the seasons. In its present position spring is shorter than summer, and autumn longer than winter; and while the solar perigee continues as it now is between the solstice of winter and the equinox of spring, the period including spring and summer will be longer than that including autumn and winter. In this century the difference is between seven and eight days. The intervals will be equal towards the year 6483; but when it passes that point the spring and summer taken together will be shorter than the period including autumn and winter. In the course of a revolution of the ellipse each season will by turns have been the longest and the shortest.

This motion of the perigee and apogee, for of course it

is common to both, is deduced from observation. If the very instant of the sun's being in apogee could be readily determined, this motion could easily be ascertained. For his place in the heavens might be determined for two successive returns to the apogee, and the difference of place would measure the motion. The variations of his apparent diameter or of his angular motion are however too slow to allow any very accurate estimation of very small differences.

§ **423.** The principle by which his apogee and perigee or the apsides of his orbit are found is very simple. We have seen that the radius vector describes equal areas in equal times. Now the only straight line which can be drawn through the focus of an ellipse so as to divide the ellipse into two equal parts is the transverse axis. If the sun be observed at any two points 180° distant from each other, he is at the two extremities of a line passing through the focus of the ellipse, and the portions of the ellipse on each side of that line must be unequal unless the line be the transverse axis, which passes through the apogee and perigee. If the portions are unequal his time of passing through them must be unequal also. If the times are equal, the instants of observation must have been when he was in apogee and perigee. If unequal, the place of the apogee and perigee may be found by calculation. The result of observation is that the apsides have a progressive motion on the ecliptic of about 11.″8 annually. The longitude of the perigee and apogee therefore increases at the rate of about 62″ annually, 11.″8 by the onward motion of those points, and 50.″1 by the retrocession of the equinox from which it is measured.

§ **424.** We now see that there are three different periods at which the sun may, in different senses, be said to return to the same position; when he returns to the same equinox at which he was before; when he returns to the same point in his orbit; and when, having been in perigee or apogee, he returns to it again; or, which is the same thing, when having been at a given distance from any of these points, he returns to the same point with respect to them. Each of these may be said to be the completion of a revolution of the sun; and a revolution of the sun is called a *year*. The year from equinox to equinox is called the equinoctial year, or sometimes the tropical year; for his time of returning from tropic to tropic, they being situations always holding the same relation to the equinox for the time being, is obviously the same as that from equinox to equinox. The year from any point in the ecliptic to the same point again is called the *sidereal year*, for the sun is then in the same position as before, with relation to the stars. The sun's angular distance from the apogee is called the true anomaly, and the period between his leaving and returning to a given situation with respect to the apogee is therefore called the *anomalistic year*.

§ **425.** It is evident that the equinoctial is the shortest, and the anomalistic the longest of these years. When the sun starts from the equinox, it is a given point of his orbit; before he returns to it, the equinox has receded on the ecliptic, and he therefore meets it again sooner than he returns to the same spot in his orbit. The effect therefore of the retrograde motion of the equinoctial point on the ecliptic is to *bring forward* the time of the equinox, (or the instant at which the sun is at the equator); and hence, as we have already mentioned, the phenomenon is known by the name of *the precession of the equinoxes*. In the mean time, however, the apogee has moved forward on the ecliptic; and the sun, therefore, after returning to the same point in his orbit where he was at the former equinox, has still a further arc to describe before he arrives at his original position with respect to the apogee, and the time of his doing so is of course later.

The *mean length of the equinoctial year* is 365d. 5h. 48m. 51.6s., (or decimally 365d.242264) of mean solar time. After this, the sun has to describe 50.″1 to return to the same point of his orbit at which he was at the commencement of the year, or to complete the sidereal year; and the *mean length of the sidereal year* is thus made 365d. 6h. 9m. 11.5s. or 365d.256383. He then has to describe a further arc of 11.″8 to arrive at his original position with respect to the apogee, and *the length of the anomalistic year* is thus made 365d. 6h. 13m. 58s.8, or 365d. 259708.

§ **426.** The lengths assigned to the equinoctial and sidereal years are only *mean* lengths; that given to the anomalistic year is a true one. We shall hereafter shew, from other considerations, that the length of the anomalistic year does not vary. For the present, we will assume that fact, and then it is obvious that the length of the equinoctial and sidereal years must continually vary; for each of these years is shorter than the anomalistic year by the time which the sun takes to describe a given angle of his orbit; in one case 62″, in the other 11.″8. Now the rate of the sun's motion is different in different parts of his orbit, faster as he is further from the apogee, slower as he approaches it; and, consequently, his times of describing these spaces of 62″ and 11.″8 continually vary as they are differently situated with respect to the apogee. The times therefore which are to be subtracted from the uniform length of the anomalistic year, to ascertain those of the equinoctial and sidereal years respectively, are themselves of variable duration; and the lengths of the equinoctial and sidereal year are necessarily so too. The variation, however, is very small, and the mean differs from the true length at any period by a very inconsiderable quantity.

§ **427.** It is obviously necessary, for many purposes, not only of chronology and history, but even of personal and domestic convenience, that we should have the means of dividing time into definite periods of considerable length; and the most obvious and natural period to adopt, is that which includes all the various operations and appearances which succeed each other in regular order, which comprehends seed-time and harvest, summer and winter.

The tropical or civil year, of 365d. 5h. 48m. 49s.7, is the time elapsed between the consecutive returns of the sun to the mean equinoxes or solstices, including all the changes of the seasons, is a natural cycle peculiarly suited for a measure of duration. It is estimated from the winter solstice, the middle of the long annual night under the north pole. But although the length of the civil year is pointed out by nature as a measure of long periods, the incommensurability that exists between the length of the day and the revolution of the sun, renders it difficult to adjust the estimation of both in whole numbers.

§ **428.** If the revolution of the sun were accomplished in 365 days, all the years would be of precisely the same number of days, and would begin and end with the sun at the same point of the ecliptic. But as the sun's revolution includes the fraction of a day, a civil year and a revolution of the sun have not the same duration. Since the fraction is nearly the fourth of a day, in four years it is nearly equal to a revolution of the sun, so that the addition of a supernumerary day every fourth year nearly compensates the difference. But in process of time further correction will be necessary, because the fraction is less than the fourth of a day. In fact, if a bissextile be suppressed at the end of three out of four centuries, the year so determined will only exceed the true year by an extremely small fraction of a day; and if, in addition to this, a bissextile be suppressed every 4,000 years, the length of the year will be nearly equal to that given by observation. Were the fraction neglected, the beginning of the year would precede that of the tropical year, so that it would retrograde through the different seasons in a period of 1,507 years.

The division of the year into months is very old and almost universal. But the period of seven days, by far the most permanent division of time, and the most ancient monument of the common origin of the human race, was used by the Brahmins in India with the same denominations employed by us, and was alike found in the calendars of the Jews, Egyptians, Arabs and Assyrians. It has survived the fall of empires, and has existed among all successive generations.

CHAPTER XIX.

PARALLAX.

Parallax defined. Horizontal Parallax. Methods of correcting for Parallax. Determination of the Moon's Parallax. Transits of Mercury and Venus. Methods of computing the Solar Parallax. Distances of the Sun and Planets. Parallax of the Fixed Stars.

§ **429.** We are now prepared to enter more particularly into the effects of parallax, both diurnal and annual, and to show the knowledge obtained by its means.

The centre of the earth is really the place to which all motions in the solar system should be referred. The centre of the sun is the point to which all out of it should be referred. Yet we are never in either of those places; we are always 4,000 miles from the one, and upwards of 95,000,000 from the other. Observations made on different parts of the surface of the earth must therefore be corrected and referred to the centre of the earth. We must always know on what part of the earth, at what period of her rotation and of her revolution, a given observation was made.

But this eccentric position, which seems so disadvantageous, is the only foundation of an accurate knowledge of the absolute dimensions of the solar system. Without it we could not ascertain the distances, and consequently the real magnitudes of the heavenly bodies.

§ **430.** The sun, moon and planets assume different positions among the fixed stars which are at an incalculable distance, when viewed by two observers in different parts of the globe. This difference of position is termed the *parallax* of the heavenly body; and when it is in the horizon of one of the observers such difference is called its *horizontal parallax.*

To an observer at A, (Fig. 7, Plate II.) a heavenly body B, will appear in the horizon, either rising or setting, and he would refer its position among the fixed stars to the point G in a circle immeasurably distant, although the limits of the diagram compel us to contract its dimensions. Another observer at A′, will have the same body, which we will suppose to be the moon, in the zenith; a line joining C, the centre of the earth, and his position, will pass through B. This observer, then, views the object as it would appear from the earth's centre, and refers it to the point E; a position distant from the former by the arc E G. Now if two observers remark the position of the moon at the same instant of time, and afterwards compare notes, the measure of this arc will be known, and consequently the value of the angle E B G, which is equal to the angle A B C, or the angle which the earth's semi-diameter would subtend when seen from B, the moon.

§ **431.** Since the triangle C A B, is right angled at A, we know A C, the earth's semi-diameter, the angles C B A and C A B; whence may be found the side C B, by the following proportion:

Sine A B C : rad. : : C A : : C B.

In the case of the moon, her mean horizontal parallax is 57′ 12″; therefore,

As sine 57′ 12″ : rad. : : 3,956 miles, the earth's semi-diameter : 237765, the moon's distance.

Her diameter may be easily found when her distance is ascertained, (Fig. 8, Plate II.): the angle E C G is that subtended by the diameter of the moon; half of this will be E C B, which is one angle of the right-angled triangle E C B, of which the base C B is known, whence the perpendicular E B may be easily found by trigonometry: this multiplied by two will give E G the diameter of the moon, which is about 2,000 miles : her mean distance is about 240,000 miles.

§ **432.** As the heavenly body rises above the horizon, its parallax will become less and less; thus, at H, (Fig. 7, Plate II.), its place, as seen from the centre, will be S; from A, it will be W; its arc of displacement S W being less than E G; when at N, Q S will be less than S W; while at M, in the zenith, its parallactic angle will vanish, and its position as seen from C, and also from A, will be I.

When the parallax of a body for one altitude is known, its parallax for any other may be found from the following proportion:

As rad. : sine of apparent zenith distance : : horizontal parallax : parallax in altitude.

For in the triangle A C H,

A C : C H : : sine A H C : sine C A H, or its supplement M A H.

Again in triangle A B C,

A C : C B : : sine A B C : sine C A B, or radius.

And since the antecedents are equal, the consequents are proportional; therefore,

Sine A H C : sine M A H : : sine A B C : rad.

Or, as rad. : sine M A H : : sine A B C : sine A H C.

Again, A C being known, and C H found, as before; in the triangle H A C given C H=C B; the angle H A C= supplement of zenith distance, whence angle A H C, which is the parallax in altitude, may be found.

The general effect of parallax causes heavenly bodies to appear nearer the horizon than their true place. Its true altitude is its observed altitude and its parallax. The horizontal parallax (Fig. 7, Plate II.) is less for a body more distant, as at D: where it is seen depressed only by the small arc F G. An exact proportion exists between the distances and horizontal parallaxes of two bodies: for in the triangle C B D, as sine B D C : sine C B D, or its supplement A B C : : C B : C D. That is, the distances of two bodies are in the inverse proportion of the sines of their horizontal parallaxes.

§ **433.** The two observers may be situated in different latitudes, but it is important that they should be under the same meridian. It is not likely that two observatories will be found situated under precisely the same meridian. Allowance must therefore be made for the change of the moon's actual zenith distance in the interval of time elapsing between its arrival on the meridians of the stations. Of course the nearer the stations are to each other in longitude, the less is this interval of time, and the smaller the correction to be made. Suppose two observers, one in north and one in south latitude, to observe on the same day the meridian altitudes of the moon's centre. Having found the zenith distances, and cleared them of the effects of refraction, if the distance of the moon were equal to that of the fixed stars, the sum of the zenith distances thus found would be precisely equal to the sum of the latitudes north and south of the places of observation. For it would be equal to the meridional distance of the stations across the equator. But the effect of parallax being in both cases to increase the apparent zenith distances, their observed sum will be greater than the sum of the latitudes, by the whole amount of the two parallaxes. This angle then is obtained by subtracting the sum of the latitudes from the sum of the zenith distances; and this once determined, the horizontal parallax is easily found by dividing the angle so determined by the sum of the sines of the two latitudes. It is curious that these latitudes are originally learned only from comparison with the stars. Thus by taking a point on the earth, and comparing it with two others, one of which is infinitely distant and the other not so, we learn the distance of the latter.

§ **434.** The moon's parallax is as useful to us as her eclipses. They teach us the globular form of the earth and moon. The variation of her parallax for the same place at different times proves to us the eccentricity of her orbit. The variation at different parts of the earth, when she is at her mean distance from them in her orbit, shows the difference in the lengths of the terrestrial radii, and thus proves that the earth is not a perfect sphere. The equatorial parallax is $\frac{1}{309}$ greater than the polar parallax. This difference, though very small has a very considerable influence upon several astronomical phenomena; for example upon the time of the occultation of a star by the moon, or even upon the occurrence of this phenomenon.

Parallax acts only in a vertical circle; it diminishes the altitude, and affects thus the right ascension, declination, latitude and longitude. This is of importance in many cases, and particularly with regard to solar eclipses. It in a measure counterbalances the effects of refraction; it depresses vertically, while refraction raises vertically.

§ **435.** Not only are the sun and moon displaced by parallax, but their apparent size is affected by it. Let us consider the distance of the moon from the centre of the earth as constant. It will then describe its daily circle round this centre, and will be nearer the observer when in the zenith than when in the horizon. If it is nearer, its apparent diameter will be greater. The increase, in passing from the horizon to the zenith is about one sixtieth, because in the passage from one of those positions to the other, the distance of the observer from the moon is diminished by a quantity equal to the radius of the earth, which is about its sixtieth part.

In the same manner, a cloud which near the horizon looked like a mere speck, when it drifts across our zenith appears a broad canopy, but subsides to its original size as it floats to a distance.

Thus a small cloud interposed between us and the sun obscures it; but an observer at a short distance refers the cloud to another place, and sees the sun also. Thus in a solar eclipse, the observer where the point of the shadow falls, loses sight of the sun entirely. To an observer on the east side of the earth, the east edge of the sun is obscured; to one on the west side, the west edge is only obscured. As the planets and the sun are much more distant from the earth than the moon is, it follows that their parallactic angles are much less than hers.

§ **436.** The sun's parallax indeed is so small that it cannot be obtained accurately by direct observation, but may be calculated from the parallax of a planet, or may be approximated in a rude way by the moon's parallax. On drawing a figure, it will be immediately seen that when the moon has completed her first quarter, the sun, the moon, and the spectator form a triangle which is right angled at the moon. Now the angle which separates the sun from the moon can be observed at the same instant; suppose it $= E$, we have,

$$\text{Distance earth and sun} = \text{distance earth and moon} \times \sec. E.$$

The exact moment when the moon is thus situated cannot be noted with accuracy, yet early observations made thus, showed that the sun was far more distant than the moon.

Kepler's discoveries, that the planets move in ellipses round the sun in the focus, that the area swept by each radius vector in a given time is a constant quantity for the same planet, and, lastly, that the squares of the periodic times are as the cubes of the mean distances, have supplied means for a much more accurate determination of the sun's parallax. Assuming these laws, the forms of the orbits of the earth and planets, and their relative distances can be determined from observation; hence, if the parallax of any one planet can be found, the parallaxes of the sun and of all the other planets can be computed. Observations of Mars, for instance, made at the Cape of Good Hope and at Greenwich, will afford a very tolerable value of his parallax, and hence of his distance. Then, as the proportion between the distances of Mars and the earth from the sun at any time is known from the form of their orbits and their periodic times, and the angle between the sun and Mars at the earth can be observed, the triangle between the sun, the earth, and Mars, can be solved, and hence the distance of the sun, and his parallax can be computed.

§ **437.** But the planets are so remote, and their parallaxes consequently so small, that they cannot be ascertained with certainty. Their parallax would scarcely be perceptible except near the horizon, and there refraction prevents any great nicety of observation.

It is only when an inferior planet passes between the earth and the sun, that very nice observations can be made both upon the planet and the sun. When Venus passes between the sun and earth, she appears like a dark round spot crossing the surface of the sun from the eastern to the western edge. Since the orbit of Venus can intersect that of the earth in only two points, the nodes, she is never seen upon the sun except when her conjunctions happen in or near the nodes of her orbit. At all other times she passes above or below the sun, and her dark side being toward the earth she is invisible. These transits take place at intervals of about eight and 113 years.

The transits of Mercury are much more frequent, but he is so much nearer to the sun than to the earth that his parallax is of little use in determining that of the sun. A small error in measuring would cause a much greater error in the snn's parallax.

§ **438.** Transits of Mercury and Venus are interesting, as proving that they are nearer to the sun than the earth is, and that they shine only by reflected light. They also show the place of the planet's nodes. How often since the creation those planets had crossed the sun's disc we know not. It was not till 1769 that a transit of Venus was observed with sufficient accuracy to fix the horizontal parallax of the sun. For the sake of observing this transit many European governments sent expeditions to various parts of the world, and the result of their observations we now consider the mean parallax of the sun. To determine the parallax with any degree of precision requires the best instruments, the most practised observers, and a combination of favorable circumstances. As a very minute error in measuring the parallactic angle at Venus would introduce an enormous error into the calculation of the distance, it is not surprising that great anxiety should have been felt to determine this angle with the utmost exactness.

§ **439.** The chief fact to be borne in mind, in considering a transit of Venus, is that the difference of the apparent beginnings and endings of the transit in different places on the earth's surface depends on the difference of the distances of the sun and Venus from the earth. If the planet were as far off as the sun, and really passed over the surface of the sun, there would be no sensible parallax in different parts of the earth. The nearer Venus is to the earth, the greater is her displacement to all observers but those situated in a right line joining the centre of Venus and the centre of the earth.

We must remember also that the apparent path of Venus depends partly on her motion, partly on the position and motion of the spectator. But as this is a complicated subject we shall consider separately the several circumstances of the transit, and give several modes in which the parallax may be found.

§ **440.** A right line passing through the centre of Venus and a given point of the earth and produced to the sun's disc, will mark the path of Venus on the sun, as seen from the given point on the earth.

When the given point is in one of the poles, there will be no parallax arising from longitude, because the pole does not move. But there will be a parallax from the latitude. The transit line seen from the pole will be parallel to that seen from the centre of the earth, though not coincident with it. When the given point is in any part of the surface whose latitude is less than 90°, there will be a parallax produced by the latitude of the place and also by its longitude. This latter parallax will alter the transit line both in position and length; and will prevent its being parallel to the central transit line, unless when the axis of Venus's orbit and that of the earth coincide, as seen from the sun; and this may not happen in many ages.

§ **441.** Let us suppose the earth perfectly at rest, without rotation or revolution, and Venus moving from west to east, with the excess of her orbitual velocity over the earth's. That part of her orbit in which she would move during her transit on the sun may be considered as a straight line; and therefore a plane may be conceived to pass through it and through the earth's centre. To every place on the earth's surface cut by this plane, Venus would be seen on the sun in the same path that she would describe as seen from the earth's centre. And therefore she would have no parallax of latitude north or south; but would have a greater or less parallax of longitude, as she is more or less distant from the meridian of each place during her transit. To all places above and below this plane she would have a parallax in latitude.

The transit of 1769 passed over the southern hemisphere of the sun's disc. Therefore a northern parallax caused her to describe a longer line on the sun than when seen from the centre of the earth. A southern parallax caused her to describe a shorter line.

To all places situated in a plane perpendicular to the orbit of Venus, prolonged to the centre of the earth, there will be no parallax in longitude. In all places eastward of this the parallax would be west of the sun ; to all places west of this the parallax would be east of the sun.

§ **442.** A favorable position for an observer is to be three hours west of the meridian which is in the plane of the earth's and the sun's centre and the pole of the orbit of Venus. After the transit has begun at the centre of the earth, it will be delayed at the western observer's place by almost the horizontal parallax of Venus. When the transit ends the observer will be three hour circles west of the meridian, and the end of the transit will be accelerated by nearly the horizontal parallax of Venus. If the observer is situated in high latitudes, being nearer the axis of the earth, the parallax from position will be less than at the equator. The parallax from rotation will be the same, for he moves with the same angular velocity.

If now we suppose another observer standing over the pole, on the opposite meridian, in such a manner that while the former moves the opposite way from Venus, he moves the same way, it is evident that to him the duration of the transit will be longer than it would at the centre of the earth. He accompanies Venus in her orbit, and thus keeps her longer between him and the sun. His position would not be all the time in the illuminated hemisphere. He would see the beginning of the transit just before sunset, and the end of it after sunrise. He must of course be so near the pole that his night lasts less than six hours, and in north or south declination according as the declination of the sun is north or south. Transits which take place in June are more favorable for observation than those in November, because the northern hemisphere offers more points of observation.

§ **443.** I shall now give two modes of finding the parallax of Venus.

In the first mode the observers are 90° apart. They each observe the time of the egress of Venus from the sun's disc. The time which elapses between these egresses is the measure of Venus's motion during this time. The arc passed through by Venus in this time is the difference between the parallax of the sun and that of Venus. The parallax of Venus may be found by the second person's ascertaining her place at the moment when he first saw her leave the sun. The former arc may be subtracted from the latter, and gives the sun's parallax.

Let D B A (Fig. 9, Plate II.), be the earth, V Venus, and T t R the western limb of the sun. To an observer at B, the point *t* at that limb will be on the meridian, its place referred to the heavens will be at E, and Venus, as it leaves the sun's disc, will appear just within it at t. But, at the same instant, to an observer at A, Venus is west of the sun, in the straight line A V F; the point *t* of the sun's limb appears at *e* in the heavens; and if Venus were then visible, she would appear at F. The angle C V A is the horizontal parallax of Venus; and is equal to the opposite angle F V E, whose measure is the arc F E. A t C is the sun's horizontal parallax, equal to the opposite angle *e* t E, whose measure is the arc *e* E; and F A *e*, is the difference between the horizontal parallaxes of Venus and the sun, and is found by observing how much earlier in absolute time Venus's total egress from the sun is, as seen from A, than as seen from B, which is the time she takes to move from *v* to V in her orbit.

§ **444.** The motion of Venus at the time of her last transit, was known to be 4′ of a degree on the sun's disc in 60′ of time, or 4″ in one minute of time.

Let us suppose then that A (Fig. 9, Plate II.) is 90° east of B, so that when it is noon at B, it will be six in the evening at A; that the egress, as seen from B, is at one minute before twelve; but that, as seen from A, it is seven minutes, thirty seconds before six. Deduct six hours for the difference of meridians of A and B, and the remainder will be 6′ 30″ for the time by which the egress of Venus on the sun at *t* is earlier as seen from A than as seen from B; which time being converted into parts of a degree is 26″, or the arc F *e* of Venus's horizontal parallax from the sun. For as one minute of time is to four seconds of a degree, so are 6½ minutes of time to 26 seconds of a degree.

§ **445.** The horizontal parallaxes of the planets are inversely as their distances from the earth's centre.

Therefore, if on the day of transit the horizontal parallax of Venus be ascertained, the sun's horizontal parallax, and consequently his distance from the earth, can be calculated.

The sun's diameter is previously known, and consequently the distance from his centre of chords of any given length. Venus displaced in opposite directions by parallax measures on the sun's disc two chords whose length is known from the duration of the transit in each place. The belt between them is twice the parallax.

Let E (Fig. 10, Plate II.) be the earth, V Venus, and S the sun, and C D the portion of Venus's relative orbit which she describes while crossing the sun's disc. Suppose A and B two spectators at opposite extremities of that diameter of the earth which is perpendicular to the ecliptic, and to avoid complicating the case, let us omit all consideration of the earth's rotation, and suppose A and B to retain that situation during the whole time of the transit. Then, at any moment when the spectator at A sees the centre of Venus projected at a on the sun's disc, he at B will see it projected at b. If then the spectator could suddenly transport himself from A to B, he would see Venus suddenly displaced on the disc from a to b; and if he had any means of noting accurately the place of the points on the disc, he might ascertain the angular measure of a b as seen from the earth.

§ **446.** Though the distance between these points cannot be immediately ascertained, the breadth of the zone included between the two apparent paths of the centre of Venus across the sun's disc may be ascertained. Each observer need only, with the utmost care and accuracy, each at his own station, notice where it enters and where it quits the sun, and what segment of the sun's disc it cuts off. This can be done with great delicacy by noting the time occupied in the whole transit. For the relative angular motion of Venus being precisely known, and her apparent path being very nearly a straight line, these times give a measure, on a very enlarged scale, of the lengths of the chords of the segments cut off. And the sun's diameter being also known with great precision, the versed sines of these chords, and therefore their difference, or the breadth of the zone required, become known. To obtain these times correctly, each observer must ascertain the instants of ingress and egress of the centre. He must note the instant when the first visible impression on the edge of the disc at P (Fig. 10, Plate II.) is produced, or the first external contact; and again, when the planet is just wholly immersed, and the broken edge of the disc just closes again at Q, or the first internal contact; and he must make the same observations at the egress at R S. The mean of the external and internal contents gives the entry and egress of the planet's centre. The especial excellence of this method consists in the nicety with which the first streak of light after the interior contact at ingress, and the last streak before the interior contact at egress, can be observed.

§ **447.** Though we have thus far been supposing the distances both of the planet and sun to be unknown, astronomers have long been acquainted, from the theory of gravity, with the proportion they bear each other. For, by comparing the rapidity of the earth in its orbit with that of Venus, the power of the sun's attraction on each may be dedueed, and thence their relative distances and likewise the proportion their distances bear each other. The distance of Venus from the sun is to its distance from the earth as 68 to 27. Since A V a, B V b, (Fig. 10, Plate II.) are straight lines, and therefore make equal angles on each side V, a b will be to A B as the distance of Venus from the sun is to its distance from the earth; as 68 : 27 : : a b : A B; a b therefore occupies on the sun's disc a space 2½ times as great as the earth's diameter, and its angular measure is about equal to 2½ times the earth's apparent diameter at the distance of the sun, or which is the same thing, to five times the sun's horizontal parallax. Having the horizontal parallax, we may easily find the distance of the sun in miles. As sine of horizontal parallax 8.6 : rad. : : 3956 miles=the earth's radius : 95,000,000=the sun's distance.

§ **448.** Knowing the distance of the earth from the sun, we may take other measurements in the system simply by observing angular distances. For instance, we may determine the distance of an inferior planet by merely measuring its angular distance from the sun at the time of its greatest elongation.

Thus if we measure the angular distance of Mercury from the sun at the time of its greatest elongation, and form a right-angled triangle by the lines joining the three bodies, we have given the distance of the earth from the sun, and the angle formed at the earth—whence may be found the distance of Mercury at the time of observation. This planet is found to vary its greatest elongation from the sun very considerably, this angle varying from 28° 48′ to 16° 12′; whence we conclude that its orbit is very elliptical. The planet Venus, on the contrary, shows but a slight deviation from a circular path; her angle of greatest elongation ranges between 47° 48′ and 44° 57′.

§ **449.** The distance of a superior planet from the sun may be found by measuring the arc through which it appears to retrograde, when in opposition to the sun.

Let X (Fig. 11, Plate II.) be the place of Mars in opposition, and y that of the earth; y v the arc described by the earth in a short period of time, for instance one day; $\chi\,\beta$ the arc described by Mars in the same time. The periodic times being known, these arcs may be found simply by dividing 360° by the number of days occupied in one revolution. Since the earth's motion is more rapid, it will pass by Mars, and Mars will appear to retrograde through the arc r d, whereas in fact he has been moving through r d′. Now in the triangle S v β, the angle S β v the difference between the heliocentric and geocentric places of Mars; the angle β S v, which is the difference between the angular advance of the earth and that of Mars for a day; and the side S v are given; whence may be found by trigonometry S β, which is the distance of Mars from the sun.

§ **450.** The whole parallax at a given distance is always equal to the angle under which the earth's diameter is viewed from that place. The more distant the observer, the smaller is the angle. All bodies within the solar system have some parallax to one another. An inhabitant of Jupiter would have a more perceptible parallax for bodies at the same distance than an inhabitant of the earth, because Jupiter would subtend to such a body a larger angle than the earth. In annual parallax Jupiter and the outer planets have, from the superior size of their orbits, a still greater advantage. Perhaps to them the parallax of the fixed stars is distinctly visible. But if the inhabitant of the earth would go beyond this system, and would displace the stars, he must seek no longer for a diurnal but an annual parallax; he must substitute 190,000,000 for 8,000 miles, and reduce observations to the centre of the orbit, not that of the earth. When the Copernican theory of placing the sun in the centre was first introduced, a strong objection to it was, the enormous distance at which it required the fixed stars to be placed. It was argued that the earth in her yearly circuit must displace them, and cause them to perform orbits parallel to the ecliptic round their true place, similar and equal to the earth's orbit round the sun. And further, that as parallax is always in a plane passing through star, sun and earth, it would diminish the angle which the sun and star subtend at the earth. But the angles remained unaltered, the circles were not described, the stars were compelled to take a position so distant that the whole orbit of the earth was from them an invisible point.

§ **451.** When the motion of the earth in an orbit was first discovered, Galileo predicted that annual parallax would reveal the distance of the fixed stars. But not the closest investigation, renewed from time to time as instruments of greater power inspired new hope, resulted in any thing but a conviction that the parallax, if any existed, was inconceivably small. The delicacy of the researches made may be judged from the discoveries elicited in the course of them. While searching for parallax, Bradley discovered nutation, and aberration of light. Sir William Herschel learned the connection and orbitual motion of the Binary Stars. But the question of the distance of the fixed stars remained unanswered for three centuries. It has lately been solved by means of more delicate instruments and a better choice of stars. There are only two circumstances to guide in the selection of stars for observation, their remarkable brilliancy or greater apparent motion. Both seem to indicate the comparative nearness of the star; and of course the nearer the star the greater the parallax.

§ **452.** Sir William Herschel first pointed out the mode of detecting parallax, which has since proved successful; namely, that which depends on the measurement of double stars. If the stars which compose a double star be at different distances from the earth, they must be differently affected by parallax, and therefore their apparent distance from each other will be altered by a change of position in the spectator. The apparent distance of two neighboring stars can be measured with great accuracy. It is more easy to determine the distance of a double star to 0.″1, than to fix an absolute place to 1″. The problem of parallax is therefore reduced to that of finding a double star in which a variation of distance is observable, and following the law which the earth's change of place requires. It was during this inquiry that Sir W. Herschel discovered that very many double stars have a relative motion, both in distance and in angular position, which proves them to be a connected system. Sir John Herschel showed that the variation produced by parallax in the angle of position of two stars is a more sensible phenomenon, and one more easily measured than the variation in distance, and he published a list of stars suitable for this research, with the times of year when the observations would show the greatest effect of parallax.

§ **453.** It has been shewn that the earth's change of place in its orbit would cause each star to have an apparent motion in an ellipse of which the major axis is parallel to the ecliptic and equal to the diameter of the earth's orbit, as seen at the distance of the star. If then the stars be a connected system, and comparatively near each other, the stars will appear to describe two equal and similar ellipses, and the line joining their apparent places will be equal and parallel to the line joining their true positions. It is therefore in vain to look for any effect of parallax in micrometrical or relative measures of distance and position in a connected system. But if one of the stars be much farther from us than the other, suppose it ten times farther off, then the apparent ellipses will continue to be similar, and similarly described; only the dimensions of that described by the more distant star will be one tenth of that described by the nearer star. Or we may suppose the more distant star to be fixed, and the nearer star to describe round its true place an ellipse of nine-tenths the actual dimensions.

§ **454.** The apparent places of the stars being similarly situated in the two ellipses, their apparent distance on the line joining these apparent places will both oscillate in angular position and fluctuate in length, thus causing an annual relative alternate movement between the stars both in position and distance which is greater, the greater the difference between the parallaxes. Thus it is not the absolute parallax of either, but the differences of their parallaxes, which is measured by this method. But when the stars are very unequal in magnitude, it is probable that the difference of their parallaxes very nearly equals the whole parallax of the nearer one.

Most of the close double stars are probably connected, and therefore unfit for the detection of parallax by comparison with one another. Yet the distance must not exceed a few seconds, for the eye, in delicate measurement, must see both stars at the same glance. The number of stars apparently double, yet so close as to admit accurate measurement, is probably small.

§ **455.** The selection of a star for observation involves many considerations. The one chosen by Mr. Bessel, 61 Cygni, is a fine double star; and one that has been ascertained to be physically double. The distance of its individuals is great, being about $16\frac{1}{4}''$. This being necessarily less than the axis of their mutual orbit, affords in itself a presumption that the star is a near one. And this presumption is increased by the unusually great proper motion of this binary system, which amounts to nearly $5''$ per annum, and which is not shared by several small surrounding stars. Moreover, the angular rotation of the two one about another has been well ascertained. Of course the proper motion and that of rotation were both allowed for in the calculation of the parallax of the central point. Of these small surrounding stars, two are very advantageously situated for comparison with either of the individuals of the binary star, or with the middle point between them. One of these (a), at a distance of 7′ 42″, is situated nearly at right angles to the direction of the double star; the other (b), at a distance of 11′ 46″, nearly in that direction. Considering a and b as fixed points, and measuring at any instant of time their distances from c, the middle point of the double star, the situation of c relative to a and b is ascertained. And if this be done at every instant, the relative place of c, or the curve described by it on the plane of the heavens with respect to the fixed base line a b, will become known.

§ **456.** Now on the hypothesis of parallax, this curve ought to be an ellipse of one certain calculable eccentricity, and no other. And its major and minor axes ought to hold with respect to the points, a, b, certain calculable positions and no others. The distances a c and b c will each of them be subject to annual increase and diminution, in a given calculable ratio the one to the other; and the maximum and minimum of the one distance a c will be nearly contemporaneous with the mean values of the other distance b c, and vice versa.

Thus we have in the first place several particulars independent of mere numerical magnitudes; and in the second place, several distinct relations a priori determined, to which those numerical values must conform, if it be true that any observed fluctuations in their distances a b and a c are really parallactic. So that if they are found in such conformity, and the above mentioned maximum and minimum do observe that interchangeable law above stated, there is accumulated a body of evidence in favor of the resulting parallax, to which no reasonable mind can refuse its belief. Mr. Bessel found that the particulars observed agreed with those calculated in a signal and satisfactory manner.

§ **457.** The distances were observed for several months, and it was found that the distance between one star and the central point diminished, while that between the other and the central point increased. The observations were corrected for the proper motion, for aberration, and even an allowance was made for the effect of the temperature on the micrometer screw. These being allowed for the distances obtained on the successive nights of observation, should have been exactly equal. This, however, was not the case. The observed and computed differences differed, and those differences seemed to follow a certain law, and it soon became manifest that the errors could be greatly reduced by admitting a certain amount of parallactic motion of the central point of the components of the double star. The parallax obtained thus was 0.″3309, which gives us the distance of the binary system of stars from our sun 670,000 × 95,000,000 of miles; a distance so great that light can traverse the interval only after a flight of nine years and three quarters, at the velocity of twelve millions of miles a minute. The diameter of the orbit described by the two stars of 61 Cygni round each other is about 50 times that of the earth's orbit, or 2½ times that of Uranus. Their periodic time is about 540 years. Such is the universe in which we exist, and which we have at length found the means to subject to measurement.

§ **458.** Once having learned the distance of a fixed star, if it were possible to measure its apparent diameter, it would be easy to ascertain its magnitude. But the fixed stars show no discs even with the highest powers. They increase in brilliancy so as to be painful to the eyes, and are surrounded by a haze or dawn like that of morning, but they remain mere points. A telescope which brings them two thousand times as near fails to give them any sensible diameter.

After years of incessant labor the parallax of the great star *α* Centauri, the third star of the heavens in brightness, has lately been ascertained to be a little more than nine tenths of a second, indicating a distance so enormous, that if our sun were so large as to fill the whole orbit of the earth, the sun when seen through a powerful telescope would have a radius of only nine tenths of a second. It is a magnificent double star; one star being of a deep and brownish orange, the other of a fine yellow color; each star being of the first magnitude. Their distance is at present (1850) about 9″, but it is rapidly diminishing, and in no great lapse of time, they will probably occult one another, their angular motion being comparatively small. Their apparent distance was formerly much greater, how much we cannot say for want of observations, but probably the major axis of their mutual orbit is short of a minute of space. They therefore afford strong indications of being very near our system. Added to which their proper motion is very considerable, and participated in by both, which proves their connection as a binary system. An additional presumption in favor of their proximity may be drawn from their situation in the nearest region of the milky way, among a great number of large stars.

The parallax of the double star α Lyrae has also been found by M. Struve of Dorpat. Its distance from the solar system is 771,400 radii of the earth's orbit, a space passed through by light in twelve years.

CHAPTER XX.

THE SUN AND PLANETS.

The Mass and Dimensions of the Sun. Its Atmosphere. Its light and Heat. The Solar Spots. Proposed explanations of their Appearance. The Sun's Rotation. The Centre of Gravity of the Solar System. Determination of the Orbits, Masses and Densities of the Planets.

§ **459.** The sun always presents a round luminous disc, and since the telescope has shown its rotation, this is a proof that its form is nearly or quite spherical. It is the only fixed star so near that we can learn its diameter and the appearance of its surface. Its mass is 800 times that of all the planets taken together. It is only 329,000 times that of the earth, though its solid contents are to those of the earth as 1,300,000 to 1. Hence it attracts the earth only 329,600 times as forcibly as the earth attracts it. In speaking of its size, we include the luminous atmospheres, which are at least several thousand miles deep. The solid nucleus of the sun may have a much greater specific gravity than has here been assigned to the whole orb. Its diameter is 111 times that of the earth. Its actual diameter is 900,000 miles. If its centre were to coincide with the centre of the earth, its volume would not only include the orbit of the moon but would extend nearly as far again. It could contain within its circumference more than 130,000 globes as large as the earth, and a thousand globes as large as Jupiter. Its surface seems to have a state of constant ebullition and commotion or rather instability. Agitated apparently by powerful causes, it often accumulates into masses like waves, whose summits now round, now in ridges, constitute bright places, and which seen through a telescope, are finely mottled, and far from uniformly bright, resembling a mackerel sky.

§ **460.** From the properties of the sun's rays it is known that this incandescent substance, which causes light and heat, is neither a solid nor a liquid, but a gas. The intensity of the sun's heat is proved by the facility with which the calorific rays traverse glass, a property which is found to belong to artificial fires in proportion to their intensity. The most vivid flames disappear, the most intensely ignited solids appear only as black spots on the disc of the sun when held between it and the eye. From this last fact it follows that the body of the sun, however dark it may appear when seen through its spots, may be in a state of complete ignition. Light and heat may arise from chemical changes taking place at the surface, from electricity, or from some cause unknown to us. The sun's direct light has been estimated to be equal to 5,563 wax candles placed at the distance of one foot from the eye. That of the moon is only equal to one candle placed at a distance of twelve feet.

§ **461.** Its heat and light are sent forth equally in all directions. When a planet is turned toward it, we become aware of these rays by their reflection; but they are as numerous throughout the whole system. We know its heat only by that intercepted by the earth, yet of that alone as much as is received in one year is sufficient to melt a stratum of ice forty-six feet deep covering the whole globe. Part of this heat is radiated back into space, but by far the greater part descends into the earth during summer, and returns thence in the course of the winter. The fixed stars are too distant to afford us sensible heat. The heat received on a given area exposed at the distance of the earth, and on an equal area at the visible surface of the sun, must be in the proportion of the area of the sky occupied by the sun's apparent disc to the whole hemisphere, or as 1 to about 300,000. A far less intensity of solar radiation collected in the focus of a burning glass suffices to dissipate gold and platina in vapor. The planets and the moon, though so near, shining only by reflected light, give no sensible heat, though chemical rays have been detected in moonlight. It would require 90,000 moons, as many as would fill the whole of our visible sky, to afford us light equal to that which we have in a cloudy day when the sun does not shine out.

§ **462.** It is now believed that the sun consists of a solid nucleus enveloped by an elastic non-luminous atmosphere, supporting luminous strata which to us appear as the disc of the sun. We infer the existence of this atmosphere from the appearances which accompany the spots seen in the sun. These spots are perfectly and intensely black, but are surrounded by a penumbra of a nearly uniform half shadow, thus presenting the appearance of a deep pit. There is no gradual melting of one shadow into the other, of spot into penumbra, penumbra into full light. The idea conveyed is more that of the successive withdrawal of veils, the partial removal of definite films, than of the melting away of a mist, or the mutual diffusion of gaseous media. The only theory which at all accounts for them supposes the black part of the spot to be the opaque body of the sun laid bare, or at least an opening, and the surrounding penumbra to be the non-luminous atmosphere. Luminous matter often is seen piled up round the edges of the spot, and often precedes the breaking out of a spot, indicating great agitation of the atmosphere.

The immense scale on which these spots take place, and the rapidity of their changes, confirm the idea that they take place in a mobile fluctuating gaseous atmosphere. A single second of angular measure seen from the earth, corresponds on the sun's disc to 465 miles. And a circle of this diameter (containing therefore nearly 220,000 square miles) is the least space which can be distinctly discerned on the sun as a visible area. A spot seen on the 29th of March, 1837, occupied an area of nearly five square minutes, equal to 3,780,000,000 square miles. The black centre of the spot, May 25, 1837, (not the tenth part of the preceding one), would have allowed the globe of the earth to drop through it, leaving a thousand miles clear of contact on each side of the tremendous gulf. That such a spot should close up in six weeks time, its borders must approach at the rate of more than 1,000 miles a day. 50 and even 150 spots have been seen at once on the sun. Sometimes several unite, and often a large one divides into several smaller ones, which soon vanish. They frequently appear in clusters, as at the time of the annular eclipse of the sun in 1836, when there were five near together.

§ **463.** These spots have a motion of their own independent of the sun's motion. These two facts have suggested an explanation of them from analogous phenomena which take place in the earth's atmosphere. The spots on the north side of the equator, move northward, and either join and disappear, or burst, one large one forming several small ones. Spots on the south side move southward, and exhibit the same changes.

On the earth hurricanes occur within the same latitudes, never cross the equator, and either gradually subside, or burst into smaller storms, and are dissipated. Hurricanes are moving cylinders of air which rotate with extreme rapidity, and are slowly translated, thus resembling the spots on the sun which have vorticose movement. These motions when left to themselves die away, the lower portions coming to rest much more speedily than the upper, by reason of the greater resistance below, and of the chief action's being above, so that their centre, like that of our water-spouts, appears to retreat upwards. In the same way the solar spots appear to fall in by the collapsing of their sides, the penumbra closing in upon the spot, and disappearing after it.

The spots also follow one another in lines parallel to the sun's equator, so that they are undoubtedly connected with the sun's rotation. It only remains therefore to inquire how such a circulation can be caused in the sun, so far as we know and understand it. It is evident this circulation on the earth depends on one part of the earth's becoming more heated than others, by exposure to the external source of heat; but the same effects would be produced if the heat of the sun's surface were equal in every part and escaped unequally.

§ **464.** Sir John Herschel finds this cause of inequality in the thickness of that atmosphere of the sun which extends beyond the luminous portion. He says that it has long been a question with astronomers, whether any such atmosphere exists. He considers the question settled in the affirmative by the rose-colored solar clouds seen in the total eclipse of July 8, 1842, which must have floated in and been sustained by an extensive transparent atmosphere. The deficiency of light at the borders of the sun's disc can only be caused by an atmosphere. To what distance this atmosphere may extend we have no means of judging accurately; but from the manner in which the dimness comes on, being by no means sudden on approaching the edge, but extending to some distance within the disc, it would seem to be considerable, not merely in absolute measure, but as an aliquot part of the sun's radius. The equatorial and polar portions of this envelope may differ in density, and thus oppose unequally the escape of the sun's heat.

§ **465.** Those spots which remain stationary on the sun's surface for a considerable time have a gradual motion across the sun's disc, making their appearance on the eastern edge of the sun, in about a fortnight, if they continue visible so long; they move toward the western edge and disappear behind it. A fortnight after, the same spot sometimes reappears on the eastern edge.

This motion of spots can only arise from the rotation of the sun and serve to mark its time of rotation. If the earth were stationary, this time would be equal to the time which elapses between two consecutive appearances of the spot on the western edge. But a correction must be introduced into the calculation, in consequence of the earth's motion, which, in the mean time, has been going on in the same direction as the motion of the sun on its axis. Thus, suppose the earth to be at E (Fig. 12, Plate II.) when a spot disappears; if the earth stood still at E, the inhabitants would again see the spot in the same place after one revolution of the sun on its axis; that is, when the spot had again arrived at b.

§ **466.** But in the mean time, the earth has advanced to D; the spot has therefore to describe the additional arc a b before it will disappear. Now the arc a b, which measures the angle a c b, is equal to the arc D E, which measures the angle D a E, or the portion of the earth's orbit which she has passed over in that time; hence, as 360°+a b, the whole space described by the spot : the whole time elapsed between the two disappearences : : 360° : the true time of the revolution of the sun on its axis, as it would have appeared to the earth had it been stationary at E. This time has been found to be 25.01154 sidereal days; two days less than it appears, owing to the earth's motion in the same direction.

These spots also prove that the sun is a spherical body; for a spot makes its appearance on the edge of the sun as a line, which gradually increases in breadth as it approaches the centre. As it passes on to the eastern edge, its diameter gradually lessens into a line before it entirely vanishes from view. Such appearances could arise only from the rotation of a spherical body.

§ **467.** The sun's spots not only prove his rotation and time of rotation, but they give us the plane of it, and show us where to draw his equator.

If the axis of rotation were perpendicular to the ecliptic, the movement of the spots would always be rectilinear, and parallel to the line which the ecliptic marks on the sun's disc.

But there are only two seasons of the year when this happens, in February and in August. Soon after this the motion of the spots becomes curvilinear, and three months after like an arc, which would have for a cord a parallel to the ecliptic. At the end of May, the convexity of this arc is toward the south; at the end of November it is toward the north. These phenomena prove that the sun's axis is inclined to the plane of the ecliptic. If this axis is so situated that at the end of February and of August it is at the edge of the visible disc of the sun, the eye of the terrestrial spectator will be in the plane of the sun's equator prolonged, and the path of the spots will be rectilinear. But three months after, the sun's pole will either be elevated above or sunk below its former position, so that all its lines of latitude will appear curvilinear. This axis is inclined to the ecliptic 7° 20′. It is directed towards a point half way between the polar star and Lyra, the plane of rotation being inclined a little more than 7° to that in which the earth revolves.

§ **468**. Besides rotation, the sun has a slight, irregular motion of revolution performed round the centre of gravity common to him and all his planets. The point to which the planets gravitate is not the centre of the sun, but the centre of gravity of the system; the sun and the planets always balance one another round this point. This causes the sun to describe an orbit about the centre of gravity of the system, which is a very complicated curve, because it results from the action of a system of bodies perpetually changing their relative positions. If all the planets were in a straight line with the sun, and on the same side with him, which, however, from the periods of the planets never can happen, the centre of the sun would be as remote as possible from the common centre of gravity. Yet this distance would be not more than .0085 of the radius vector of the earth. We may form a better idea of the magnitude of this orbit by comparing it with that of the moon. A body revolving round the sun in contact with his surface, must be nearly twice as remote from his centre of gravity as the moon is from the earth, and the sun in his revolution round the common centre of gravity of the system, must, when most remote, be four times the distance of the moon from the earth.

§ **469.** Before speaking of the physical constitution of the planets, we shall introduce a few remarks on the nature of the orbits which they describe, and of the methods of ascertaining their masses and density. For the complete determination of elliptic motion, the nature and position of their orbits must be observed. This depends on six quantities called the elements of the orbit, the modes of obtaining which, when not learned by direct observation, have been already shown. These are, the length of the major axis, and the eccentricity, which determine the form of the orbit; the longitude of the perihelion; the inclination of the orbit to the plane of the ecliptic, and the longitude of its ascending node. These give the position of the orbit in space; but the periodic time, and the longitude of the planet at a given instant, called the longitude of the epoch, are necessary for finding the place of the body in its orbit at all times.

The only one of these quantities which is invariable in each orbit is the length of the major axis, which equals the earth's greatest and least distance from the sun. The variations of the other elements are explained in the proper connection.

§ **470.** The periodic time of a planet is the interval between its quitting a node and returning to it again. Its nodes are the points where its orbit intersects the ecliptic. The ascending node is that point in the ecliptic through which the planet passes in going from the southern to the northern hemisphere. The descending node is a point in the plane of the ecliptic diametrically opposite to the former, through which the planet descends in passing from the northern to the southern hemisphere.

The passage of a planet through its node is seen when it actually occurs, and in its true place, whether the planet's motion at the moment appear to be slow or swift, direct or retrograde. It is easy to ascertain by observation the time when a planet crosses the ecliptic. Its right ascensions and declinations are converted into longitudes and latitudes, and the change from north to south latitude shows in what day the transition took place; while a simple proportion, grounded on the observed rate of its motion in latitude in the interval, suffices to fix the precise hour and minute of its arrival on the ecliptic. Suppose, by two observations 24 hours apart, it had been found to be 3° N. L., and 5° S. L. As 8° : 24h. : : 3° : 9h. : : 5° : 15h. It crosses the ecliptic at nine o'clock. This having been done for several transitions, and the dates having been thereby fixed, the interval of time is found to be always the same. This periodic time is the same, whether the body moves in a circular or elliptic orbit, provided only that the mean distance or half the major axis of the orbit remains the same. In a circle the motion is uniform, in an ellipse it varies, but in both the planet arrives at the extremities of the major axis in the same time.

§ **471.** Since the position of the earth and its observed motion have no effect on the time during which a planet appears in the ecliptic, the sidereal periods of the planets may be obtained with accuracy by thus noting their passages through the nodes of their orbit. A very slight retreat of the nodes must however be allowed for. The synodical revolution brings a planet to the same angular distance when viewed from the earth. It differs from the sidereal period, which brings the planet to the same situation as regards the sun. The sidereal is uninfluenced by the motion of the earth.

Mercury's sidereal period is nearly 88 days, his synodical is 116; that of Venus is less than 225 days, while her synodical revolution occupies about 584 days. The planet must not only perform its own sidereal revolution, but it must move on to compensate for the arc the earth has gone through in the same time. This arc is described by the earth with her velocity in the same time that the planet with its velocity describes that arc plus 360°. If the planet be a superior one we must reverse the calculation. The earth will describe 360+a certain arc, while the planet describes a similar arc. Let V and v be the mean angular velocity, x the arc of excess; then V : v : : 1+x : x; and V—v : v : : 1 : x, whence x is found, and $\frac{x}{v}$ = the time of describing x, or the difference between the sidereal and synodical periods.

§ **472.** To ascertain the relative masses of two bodies, we need only compare the force they exercise on bodies revolving round them. We know that the moon descends through $16\frac{1}{12}$ feet in a minute of time, owing to the earth's attraction. We know that the earth descends through 2.2 this space in a minute of time owing to the sun's attraction. But we know the distance of the moon from the earth, and of the earth from the sun. The latter distance is four hundred times the former.

The whole attraction is in a ratio made up of the ratios of the masses directly, and of the squares of the distances inversely. Thus putting F for the whole attraction of the sun, and f for the whole attraction of the earth, D and d for the distances from the sun and from the earth, we have

$$f : F :: m : M :: D^2 : d^2$$
$$f : F :: m d^2 : M D^2$$
$$f d^2 : F D^2 :: m : M.$$
$$m : M :: 1 \times 1^2 : 2.2 \times 400^2 :: 1 : 352000.$$

By marking the deflection of one of the satellites of those planets which are provided with them, and comparing it with the deflection of the moon, the comparative masses of the earth and that planet may be found. The masses of those planets which have no satellites are proved from their perturbations. We can observe how much it influences another known body at a given distance, and how much it is influenced by that body.

§ **473.** The earth's real diameter and a planet's apparent diameter and distance being known, the planet's real diameter and solid contents or size can be calculated. Its size and mass being known, its density may be compared with that of other bodies.

If we know the density of one planet we may find that of another. Suppose two planets, A and B are found to operate on a third by attractions proportionate to the numbers 7 and 2, at distances which are to each other as 4 to 3; let, moreover, their diameters be as 3 to 2.

Let d stand for the density of A, and D for that of B; then the attraction of A and B on the third body will be directly as their masses, and inversely as their distances squared. Hence, remembering that similar solids are to each other as the cubes of their diameters, if homogeneous, or in the compound ratio of their cubes and their densities, if their densities differ, the attraction of A will $=\frac{3^3 \times d}{16}$ and that of B $=\frac{2^3 \times D}{9}$.

But by observation the attractions are found to be as 7 to 2; therefore,

$$3^3 \times d \times 9 : 2^3 \times D \times 16 :: 7 : 2; \text{ or}$$
$$486\ d = 896\ D, \text{ that is,}$$
$$d : D :: 896 : 486.$$

CHAPTER XXI.

THE PLANETS—(CONTINUED.)

Mercury. Its rare visibility. Its Phases. Venus. Its brilliancy. Rotation and Atmosphere. Mars. Its Polar Spots. The Asteroids. Conjecture as to their Origin. Jupiter. Its Belts and Rotation. The Satellites of Jupiter. Theory of their Motion. Saturn. Its Rings and Satellites. Uranus. Neptune.

Mercury.

§ **474.** Mercury is the nearest planet to the sun; it appears after sunset in the west with a small but very brilliant disc. As long as it is east of the sun it remains visible after him in the evening. As soon as in its orbit it becomes west of the sun, it appears before the sun in the morning, withdraws more and more from him till it is 28° distant, then draws nearer daily, till at last it disappears, to become again an evening star. The brief duration of its appearance each night is owing to its closeness to the sun, from which it is only 37,000,000 miles distant. It is .06 the size of the earth, has a day of 24 hours, and a year of 87 days, and travels 111,000 miles an hour. Its orbit, which is always included within that of the earth, is very much inclined to the plane of its equator, and makes with the plane of the ecliptic an angle of 7°.

§ **475.** Sometimes when Mercury plunges into the sun's rays, he is seen like a black spot traversing the sun's disc; this is called the transit or passage of Mercury. His dark appearance during these transits proves that he borrows his light from the sun. His transits make it easy to determine his inclination to the ecliptic, and consequently his nodes. The line he describes in passing over the sun is sufficient to determine these. .His transits always take place either in May or November, but much more frequently in the latter month, because his orbit being quite eccentric he is much nearer the sun in our winter than in our spring. The sun forms the base of a cone of luminous rays passing to the eye of the observer, and the nearer the base Mercury passes the more probability is there of his intercepting the rays.

Mercury is of a spherical form, and exhibits phases like the moon. One horn of its crescent has been seen indented, and this proves the irregularity of its surface, and likewise gives the time of its rotation. It is caused by some mountain on its surface, which in some positions screens from our view some of the points illumined by the sun.

No spots have been seen on his surface, and nothing is known of his atmosphere, except that it contains clouds. It has a pale silvery light, and a slight bluish tint. It probably received its name from its swiftness, in which it resembles the messenger of the gods. For its motion from point to point of space is more rapid than that of any other planet.

§ **476.** The sun transmits to it seven times the heat of our torrid zone. The mean heat must be above that of boiling quicksilver, and water would boil even at his poles unless the sun's rays are modified by an atmosphere peculiar to the planet.

When Mercury and Venus are nearest the earth, or in their inferior conjunction, their dark side is toward it, and unless they pass over the sun's disc they are invisible. As they revolve in different planes from the earth, transits can only take place when they are in or near their nodes at the time of their inferior conjunction.

As they move forward in their orbits they gradually show a portion of their enlightened surface till they reach their superior conjunction, when the whole of the bright side is turned toward the earth. After which the illuminated surface decreases in size as they return toward the earth. The period of greatest brilliancy of course cannot be when they are nearest the earth, neither is it when they are most distant; but it is at a point of their orbits in which the smaller extent of the illuminated surface is more than compensated by the position of the planet nearer to the earth.

Venus.

§ **477.** Venus is the most beautiful of the planets, and for this reason received the name she bears. She is a morning star from her inferior to her superior conjunction, and an evening star from her superior to her inferior conjunction.

She is never more than 48° distant from the sun. She appears much larger at one time than at another, because her distance from the earth varies so much, and is most brilliant when in her quarters once in eighteen months, being then seen by day-light, though not very distinctly. She returns to her most brilliant position once in eight years, owing to the ratio of her periodic time to that of the earth. She is only visible for three or four hours in the morning or in the evening, according as she is west or east of the sun, being alternately before or after him for about 290 days, and appearing sooner than she otherwise would owing to her angular velocity being greater than that of the earth.

When a morning star, she rises before the sun, and is seen from the earth in the form of a handsome crescent, with its convex side turned eastward toward him. When an evening star, she follows the sun, and becomes visible after he has set, having the convex side of her crescent turned toward him.

Mercury is her morning and evening star, as she is ours.

§ **478.** The mean distance of Venus from the sun is 70,000,000 miles. Her size compared to that of the earth is $\frac{93}{100}$. Her day is 23 hours long, and her year includes 224 of our days, or 234 of her own. Her orbit is inclined 3° to the the ecliptic, and her axis of rotation is inclined to her orbit 75°, that is 51° more than the earth's axis is inclined to the ecliptic. The greatest declination of the sun on each side of her equator is 75°. Her tropics are 15° from her poles, and her polar circles at the same distance from her equator. She has therefore at her equator two severe winters and two summers in a year. The north pole of her axis is inclined toward Aquarius, the earth's toward Cancer, consequently the northern hemisphere of Venus has summer in the sign in which our earth has winter, and vice versa.

The period of the rotation of Venus is determined like that of Mercury, by the interval of time between two successive appearances of the truncated horns of her crescent. Not only is her crescent truncated at each end, but a little bright peak is visible at some distance from the illuminated surface.

§ **479.** Venus has a larger atmosphere and more marked clouds than Mercury. The existence of the atmosphere is known not only by the clouds but by its refraction. It causes the illuminated portion of her surface to look larger than it otherwise would. This dense atmosphere may perhaps cause the white and silvery appearance and the brightness by which Venus is distinguished from all other planets.

The existence of a satellite to Venus is a question which has been very much discussed. We are so unfavorably placed for seeing a satellite of hers that she may have one undiscovered by us. Its enlightened side can never be fully turned toward us but when Venus is in her superior conjunction 160,000,000 miles distant from us, and Venus then appears but little larger than an ordinary star. When she is between us and the sun, her moon would have its dark side toward us, and we could not see it any more than our own moon at time of change. It is however improbable that Venus has a moon, as modern instruments would have revealed one not extremely minute.

Mars.

§ **480.** This planet lies next outside of our earth. Its mean distance from the sun is 146,000,000 of miles. Its distance from the earth is so variable that its diameter appears at some times five times as large as at others. It moves at the rate of 55,000 miles an hour, and performs its annual circuit in an eccentric ellipse in 686 of our days. By the spots on its disc it is found that it rotates on its axis in 24 hours. Its axis of rotation is inclined to that of its orbit 30°, and the inclination of its orbit to the ecliptic is 1° 51′. Its equatorial is to its polar diameter as 16 to 15. Its diameter is about 4,100 miles.

Mars is known in the heavens by its red dusky appearance. It has a very thin atmosphere, which allows the red body of the planet to be perceived with its unvarying marks. The outlines of what may be land and sea are distinctly visible, the continents being of a ruddy color, probably belonging to the soil, and the seas consequently appearing greenish. These spots are not always to be seen, owing perhaps to clouds, but when visible they are of the same figure and position.

§ **481.** The polar regions of Mars shine with a brilliancy so superior to the rest of his disc that they appear like segments from a larger globe. As each pole appears after a long winter this lustre is observed, and is attributed to reflection from the snow and ice collected during a winter night of twelve months' duration. The brightness gradually disappears after exposure to the sun, and the pole which has been twelve months under the sun's rays is nowise distinguishable from the rest of the planet.

Mars is the only superior planet which exhibits any perceptible phase; it never appears as a crescent, but has sometimes a moderately gibbous appearance, the enlightened portion of the disc being never less than seven-eighths of the whole. By means of this gibbosity the proportional distances of Mars from the sun, and the earth from the sun, may be ascertained. Let E (Fig. 13, Plate II.) be the earth at its apparent greatest elongation from the sun S, as seen from Mars M. In this position the angle S M E is at its maximum. S M bisects the illuminated hemisphere, E M bisects the gibbous part; the diameter of the illuminated surface gives us S M E. This angle being found, the proportion of S E to S M may be found, and it appears that S E (the distance of the earth) is $\frac{2}{3}$ of S M (the distance of Mars from the sun).

The Asteroids.

§ 482. The regularity of the intervals between the other planets, and the distance between Mars and Jupiter, first made it suspected that there might be another planet between these. Four of them, Ceres, Pallas, Juno, and Vesta, were discovered near the beginning of the present century, and since 1845, five other planets, Astraea, Iris, Hebe, Flora and Metis, have been added to the group. These are sometimes called the ultra zodiacal planets, because their orbits are so much inclined to the ecliptic as to pass beyond the zodiac. This circumstance and their small size have given rise to the conjecture that they may be the fragments of some larger planet which once existed at this distance from the sun. It seems more probable that they took their present form at the same time with the other members of the solar system. They seem a connecting link between the planets and the zone of meteoric bodies which we have before described. They may thus be regarded as the smallest of the planets, or the largest of the planetoids.

Owing to their small size and their distance from us

very little is known of them. They are never visible to the naked eye, and through the telescope appear like small nebulous stars. Some of them show a disc so that their size is known, others have as yet shewn no disc. Their diameters vary in size from $\frac{1}{30}$ to $\frac{1}{100}$ that of the earth, or even less than this. Their orbits vary in inclination to one another as much as 8°.

§ **483.** Ceres was the first discovered. It revolves in 4½ years, in an orbit inclined 10° to the ecliptic. It is 264 millions of miles from the sun, and appears like a nebulous star surrounded by very variable mists.

The orbit of Pallas is extremely elongated, and its inclination to the ecliptic is 34°, greater than that of any other planet. Its distance is 267,000,000 miles. It is of a whitish color and nebulous appearance, indicating an extensive and vaporous atmosphere, little repressed and condensed by the attraction of so small a mass. Its diameter does not much exceed 79 miles, so that a steam-carriage might go round it in a few hours.

Juno has a year equal to 4½ of ours, and a distance of 256,000,000 miles. Its orbit is inclined 23°.

Vesta is smaller than those previously discovered. It has about the same surface as the kingdom of Spain. It is only 225,000,000 miles from the sun; its orbit is inclined 7°; its year is 3¼.

Astraea differs but little from Juno in size, mean distance, and periodic time. It has as yet shown no disc.

Flora equals a star of the eighth or ninth magnitude, and shines with a bluish light.

Jupiter.

§ **484.** Jupiter is the largest of the planets, and next to Venus the most brilliant. Jupiter's disc is crossed in one direction by dark bands or belts, which vary in breadth and situation. These bands have the appearance of strings of clouds. Between the bands, in the dark spaces, bright spots, on the surface of the planet, are visible. These clouds probably owe their form to currents analagous to our trade winds. Such currents must be much stronger in Jupiter than in the earth, owing to Jupiter's greater rapidity of rotation, particularly about the equator. Jupiter has probably very little change of seasons for his orbit is very slightly inclined to the plane of his rotation. His day, we have already said, is very short, only 10 hours in length, while his year contains 10,000 of his days, and 4,332 of ours. His diameter is 89,000 miles.

§ **485.** *" The changes which take place in the planetary system are exhibited on a smaller scale by Jupiter and his satellites; and, as the period requisite for the developement of the inequalities of these moons only extends to a few centuries, it may be regarded as an epitome of that grand cycle which will not be accomplished by the planets in myriads of ages. The revolutions of the satellites about Jupiter are precisely similar to those of the planets about the sun: it is true they are disturbed by the sun, but his distance is so great that their motions are nearly the same as if they were not under his influence. The satellites, like the planets, were probably projected in elliptical orbits; but, as the masses of the satellites are nearly one hundred thousand times less than that of Jupiter; and as the compression of Jupiter's spheroid is so great, in consequence of his rapid rotation, that his equatorial diameter exceeds his polar diameter by no less than six thousand miles; the immense quantity of prominent matter at his equator must soon have given the circular form observed in the orbits of the first and second satellites, which his superior attraction will always maintain. The third and fourth satellites being farther removed from his influence, revolve in orbits with a very small eccentricity. And although the first two sensibly move in circles, their orbits acquire a small ellipticity, from the disturbances they experience.

§ **486.** It has been stated, that the attraction of a sphere on an exterior body is the same as if its mass were united in one particle in its centre of gravity, and therefore inversely as the square of the distance. In a spheroid, however, there is an additional force arising from the bulging mass at its equator, which acts as a disturbing force. One effect of this disturbing force in the spheroid

* From Mrs. Somerville's Connection of the Physical Sciences.

of Jupiter is, to occasion a direct motion in the greater axes of the orbits of all his satellites, which is more rapid the nearer the satellite is to the planet, and very much greater than that part of their motion which arises from the disturbing action of the sun. The same cause occasions the orbits of the satellites to remain nearly in the plane of Jupiter's equator, on account of which the satellites are always seen nearly in the same line; and the powerful action of that quantity of prominent matter, is the reason why the motions of the nodes of these small bodies is so much more rapid than those of the planet.

§ **487.** The nodes of the fourth satellite accomplish a tropical revolution in 531 years; while those of Jupiter's orbit require no less than 36,261 years; a proof of the reciprocal attraction between each particle of Jupiter's equator and of the satellites. In fact, if the satellites moved exactly in the plane of Jupiter's equator, they would not be pulled out of that plane, because his attraction would be equal on both sides of it. But, as their orbits have a small inclination to the plane of the planet's equator, there is a want of symmetry, and the action of the protuberant matter tends to make the nodes regress by pulling the satellites above or below the planes of their orbits; an action which is so great on the interior satellites, that the motions of their nodes are nearly the same as if no other disturbing force existed.

The orbits of the satellites do not retain a permanent inclination, either to the plane of Jupiter's equator, or to that of his orbit, but to certain planes passing between the two, and through their intersection. These have a greater inclination to his equator the farther the satellite is removed, owing to the influence of Jupiter's compression; and they have a slow motion corresponding to secular variations in the planes of Jupiter's orbit and equator.

§ **488.** The satellites are not only subject to periodic and secular inequalities from their mutual attraction, similar to those which affect the motions and orbits of the planets, but also to others peculiar to themselves. Of the periodic inequalities, arising from their mutual attraction, the most remarkable take place in the angular motions of the three nearest to Jupiter, the second of which receives from the first a perturbation similar to that which it produces in the third; and it experiences from the third a perturbation similar to that which it communicates to the first. In the eclipses these two inequalities are combined into one, whose period is 437—659 days. The variations peculiar to the satellites, arise from the secular inequalities occasioned by the action of the planets in the form and position of Jupiter's orbit, and from the displacement of his equator.

§ **489.** It is obvious that whatever alters the relative positions of the sun, Jupiter, and his satellites, must occasion a change in the directions and intensities of the forces, which will affect the motions and orbits of the satellites. For this reason the secular variations in the eccentricity of Jupiter's orbit occasion secular inequalities in the mean motions of the satellites, and in the motions of the nodes and apsides of their orbits. The displacement of the orbit of Jupiter, and the variation in the position of his equator, also affect these small bodies. The plane of Jupiter's equator is inclined to the plane of his orbit at an angle of 3° 5′ 30″, so that the action of the sun and of the satellites themselves produces a nutation and precession in his equator precisely similar to that which takes place in the rotation of the earth, from the action of the sun and moon. Hence the protuberant matter at Jupiter's equator is continually changing its position with regard to the satellites, and producing corresponding nutations in their motions. And, as the cause must be proportional to the effect, these inequalities afford the means, not only of ascertaining the compression of Jupiter's spheroid, but they prove that his mass is not homogeneous. Although the apparent diameters of the satellites are too small to be measured, yet their perturbations give the values of their masses with considerable accuracy—a striking proof of the power of analysis.

§ **490.** A singular law obtains among the mean motions and mean longitudes of three satellites. It appears from observation that the mean motion of the first satellite, plus twice that of the third, is equal to three times that of the second; and that the mean longitude of the first satellite, minus three times that of the second, plus twice that of the third, is always equal to two right angles. It is proved by theory, that if these relations had only been approximate when the satellites were first launched into space, their mutual attractions would have established and maintained them, notwithstanding the secular inequalities to which they are liable. They extend to the synodic motions of the satellites; consequently they affect their eclipses, and have a very great influence on their whole theory. The satellites move so nearly in the plane of Jupiter's equator, which has a very small inclination to his orbit, that the first three are eclipsed at each revolution by the shadow of the planet: the fourth satellite is not eclipsed so frequently as the others. The eclipses take place close to the disc of Jupiter when he is near opposition; but at times his shadow is so projected with regard to the earth, that the third and fourth satellites vanish and reappear on the same side of the disc. These eclipses are in all respects similar to those of the moon; but, occasionally, the satellites eclipse Jupiter, sometimes passing like obscure spots across his surface, resembling annular eclipses of the sun, and sometimes like a bright spot traversing one of his dark belts. Before opposition, the shadow of the satellite, like a round black spot, precedes its passage over the disc of the planet, and after opposition, the shadow follows the satellite.

§ **491.** In consequence of the relations already mentioned in the mean motions and mean longitudes of the first three satellites, they never can be all eclipsed at the same time. For when the second and third are in one direction, the first is in the opposite direction; consequently, when the first is eclipsed, the other two must be between the sun and Jupiter. The eclipses of Jupiter's satellites have been the means of a discovery which, though not immediately applicable to the wants of man, unfolds one of the properties of light—that medium without whose cheering influence all the beauties of creation would have been to us a blank. It is observed, that those eclipses of the first satellite, which happen when Jupiter is near conjunction, are later by 16′ 26″.6 than those which take place when the planet is in opposition. As Jupiter is nearer to us when in opposition by the whole breadth of the earth's orbit than when in conjunction, this circumstance is attributed to the time employed by the rays of light in crossing the earth's orbit, a distance of about 190,000,000 miles; whence it is estimated that light travels at the rate of 190,000 miles in one second. Such is its velocity, that the earth, moving at the rate of nineteen miles in a second, would take two months to pass through a distance which a ray of light would dart over in eight minutes.

§ **492.** The velocity of light deduced from the observed aberration of the fixed stars, perfectly corresponds with that given by the eclipses of the first satellite. The same result, obtained from sources so different, leaves not a doubt of the truth. Many such beautiful coincidences, derived from circumstances apparently the most unpromising and dissimilar, occur in physical astronomy, and prove connections, which we might otherwise be unable to trace. The identity of the velocity of light, at the distance of Jupiter, and on the earth's surface, shows that its velocity is uniform; and if light consists in the vibrations of an elastic fluid or ether filling space, an hypothesis which accords best with observed phenomena, the uniformity of its velocity shows that the density of the fluid throughout the whole extent of the solar system must be proportional to its elasticity."

Saturn.

§ **493.** A still more wonderful mechanism is displayed in Saturn, the planet next in order of remoteness to Jupiter, and not much inferior to him in magnitude, being 79,000 miles in diameter, and 1,000 times the bulk of the earth. Seen by the naked eye, it is a star of a dull lustre, and as its motion is very slow, scarcely distinguishable from a fixed star. At a distance of 915,000,000 of miles from the sun, it moves in an orbit inclined only 2° to the ecliptic. Its revolution occupies twenty-four years, its rotation about ten hours. The flattening of its poles amounts to one twelfth of its diameter. It is surrounded by a ring which revolves nearly in the plane of its equator. This ring is about 60,000 miles broad, and 100 thick, and is about 19,000 miles distant from the planet. When this ring was first seen only the ends of it were visible, looking like the ears of a jar. It was said that Saturn was old and had two helpers to uphold him. As the planet passed on in its orbit, the ring widened out, and showed itself circular. It was then thought to be a necklace of moons, or two moons connected by a line of light. Afterward a dark stripe, concentric with the circumference of the ring, was distinctly visible, and was thought to be a deep valley. Herschel saw in this dark interval a shining spot, unlike a mountain, and at last detected it to be a star, and knew that the ring was composed of two parts.

§ **494.** It has been thought that it consists of many concentric rings, which revolve with a rapidity sufficient to counteract the attraction of Saturn. They rotate from west to east, and in the same time a satellite at an equal distance would require. That the rings are solid and opaque is shown by their casting their shadow on the body of the planet when they are toward the sun, and by their receiving the shadow of the planet on that part of them which is remote from the sun.

Since the rings always remain at right angles to the axis of rotation, and that is inclined to the ecliptic 28°, the rings must present to an observer on the earth a great variety of appearances. They must vary from a long ellipse to a mere line. When the plane of the rings passes through the sun's centre, only the edge of the rings is illuminated; when their plane passes through the earth's centre, they appear as a very fine line drawn across the disc and projecting out on each side. The rings disappear to common telescopes once in fifteen years, but the disappearance is generally double, the earth passing twice through the plane of the rings before they are carried past our orbit by the slow motion of Saturn. Then the two inner satellites appear like beads threading the slender line of light to which the rings are reduced. They separate from them at each end while turning round on their orbits. They move close to the edge of the rings, and their orbits never deviate much from their plane. To those regions of the planet which lie above the enlightened side of the rings they must appear like wide arches spanning the heavens from horizon to horizon, while the regions beneath their shade suffer an eclipse of fifteen years. Each side of these rings has fifteen years of light and fifteen of darkness. The mass of the rings is thought to be $\frac{1}{118}$ part of that of the planet. It is thought that they are not perfectly concentric with the planet, but, owing to some inequality in their thickness or their density, revolve round a centre of gravity not the same as that of the planet. If they were concentric with the body of Saturn, any disturbing cause would increase their oscillations till their equilibrium would be overthrown; as it is, the heavier parts have sufficient momentum to overcome minute disturbing forces—such as the attraction of satellites.

§ **495.** Saturn has eight satellites. The most distant but one is the largest, and is nearly of the size of Mars. The orbit of the outer satellite is inclined 30° to the plane of the ring, while the orbits of the others nearly coincide with it, and are by the attraction of the equatorial parts made circular. The outer one exhibits periodical changes, which prove that like our moon it rotates and revolves in the same time. The two interior satellites, and that recently discovered, are very minute, and can scarcely be seen except when the ring disappears. Their periods of revolution vary from twenty-two hours to seventy-nine days.

If the inhabitants of Saturn and Jupiter have such eyes as ours, unassisted by instruments, Jupiter is the only planet which can be seen from Saturn, and Saturn the only one which can be seen from Jupiter. So that the inhabitants of these planets must have much better sight than we, or have equally good instruments, to find out that there is such a body as the earth in the universe. For the earth is but little larger, seen from Jupiter, than his outer moon is seen from the earth; and if his large body had not first attracted our sight and caused us to view him through a telescope, we should probably never have known the existence of his moons or those of Saturn.

Uranus.

§ **496.** Sir William Herschel, looking through his seven-feet telescope at the stars in the feet of Gemini, saw a little star rather different from others of the same light, and apparently larger, which he suspected of being a comet. He looked at it with a power of 932, and found its diameter still increasing. He compared it with many little stars, and observed its position with regard to them. Two days after he was assured it was no star, because it had changed its place. By degrees it was found that its diameter did not change sensibly, that its orbit was almost round. After a year's observation it was found that its revolution occupied about eighty-three years. It was recognized as a star which in different catalogues had been set down successively in two different places. Its distance from the sun is 1,840,000,000 of miles. Its orbit almost coincides with the ecliptic. Its time of rotation is not known; nor its compression; but the orbits of its satellites are nearly perpendicular to the plane of the ecliptic, and by analogy they should be in the plane of its equator. Of the physical constitution of Uranus nothing is known. The earth cannot be visible, even as a telescopic body, to an object so remote. It can be seen from the earth only by excellent eyes, and its satellites only by an instrument far better than is commonly met with in observatories; their number is uncertain.

Neptune.

§ **497.** Of this planet, owing to its remoteness and recent discovery, little is known. It shines with a bluish light, and is accompanied by one and perhaps more satellites. Its mass is about $\frac{1}{20000}$ of that of the sun. The extraordinary circumstance of its having been discovered in consequence of the predictions of two eminent astronomers, gives to this planet a very prominent place in astronomical history.

CHAPTER XXII.

THE MOON.

Size and Mass of the Moon. Its Distance and Period. Revolution of the Nodes of the Lunar Orbit. Appearance of the Moon. Libration. Phases of the Moon. The Harvest Moon. The Lunar Theory. Action of the Sun. Evection. Variation. Annual Equation. Action of the Planets. Acceleration of the Mean Motion.

§ **498.** The mass of the moon is determined from several sources; from her action on the terrestrial equator, which causes nutation in the axis of rotation; from an inequality she produces in the sun's longitude; and from her action on the tides: it appears to be about one eightieth part of that of the earth. Since her volume is nearly one fiftieth that of the earth, her density is two thirds that of the earth. Her form is slightly spheroidal. Her diameter is 2,160 miles, and her average distance from the centre of the earth but 237,360 miles. She completes the circuit of the heavens in 27 days, 7 hours, 43 minutes, moving at the rate of 2,000 miles an hour, in an orbit whose eccentricity is about 12,985 miles. By observation of her parallax it is found that her mean distance is about sixty times the radius of the earth. Her greatest distance is 64, her least 56 radii of the earth, quantities which are to each other as 8 to 7, and which give a much greater eccentricity than that of the solar ellipse. Her greatest apparent diameter is 33′ 31″, her least 29′ 22″. Beside the variation in her diameter, owing to the ellipticity of her orbit, there is a slighter one owing to parallax. When the moon is in the zenith, she is nearer to an observer by the radius of the earth, or one sixtieth of her whole distance, than when in the horizon. Her diameter is accordingly 30″ larger than when in the horizon. Her orbit is inclined to the ecliptic a little more than 5°; this inclination varies; but it never falls short of 5°, nor exceeds 5° 18′. When she crosses north of the ecliptic she is in her ascending node, when she passes south of it she is in her descending node. Her nodes retreat on the ecliptic 19° 21′ a year, and complete the circuit of the heavens in 6,793d. 10h. 6′ 30″. The points in the orbit in which the moon is nearest to and farthest from the earth, are called respectively its apogee and perigee; the line joining them is called the line of apsides. This line advances at the mean rate of 40° 40′ 32″ every year, and completes the circuit of the heavens in 3,232d. 13h. 56′ 17″.

The periodic time of the moon, or that occupied in making a complete revolution round the earth, or to the same star, occupies 27d. 7h. 43′ 12″. Her synodic period or time of being again in the same direction with the sun, occupies 29d. 12h. 46′ 3″.

§ **499.** Either of these being known, the other may be computed from it. Thus let s represent the synodic period of the moon, p her periodic time, P the period of a revolution of the sun; and let A represent the angle through which the sun has moved before the moon overtakes him; and also let us suppose for the present that the angular motions both of the sun and moon are uniform. If this be the case, as the sun moves through 360° in P, he will move through $360°\frac{s}{P}$ in the time s, or in the synodic period of the moon; or the angle A will be $360°\frac{s}{P}$. But the moon, before she overtakes the sun, will have moved through $360°+a$, or $360°+360°\frac{s}{P}$; and as she takes the time p to move through 360°, she will take the time $\frac{360°+a}{360°}$ p to move through $360°+A$; or $s=\frac{360°+360°\frac{s}{P}}{360°}$ p, or $Ps=Pp+sp$, and hence $p=\frac{Ps}{P+s}$ or $s=\frac{Pp}{P-p}$; equations from which, if either quantity s or p be known, the other may be computed; for P, the length of the year is known. The angular motions both of the sun and moon being variable, this is not an accurate method of estimating these elements. It gives however their mean value.

The synodic period may be best learned from eclipses of the moon. The middle of such an eclipse is very near the time in which the earth is directly between the sun and moon, and that exact time may be computed from observations made of the eclipse. From one eclipse to another, happening under similar circumstances, must be an exact number of synodic periods of the moon. The number of such periods which have elapsed is known, and the whole interval between the two eclipses, divided by this, gives the exact length of the mean synodical period. The position of the moon's apogee in the two eclipses, and the season of the year in which they are observed, introduces some slight incorrectness, but divided among so many months, the error is unimportant. From the synodical period thus found, subtract its excess over the sidereal period. This excess is the arc the sun passes through in 29.53 days, or 29°.1. The moon, moving at the rate of 13° 17′ a day, will describe this arc in 2.21 days.

The moon's year includes thirteen of her synodic days, and each day has a fortnight's light and a fortnight's darkness.

§ **500.** The moon's disc exhibits to the naked eye numerous irregularities; observed through a telescope it is covered with crater-like hollows and with steep jagged hills. Its face is as well known to astronomers as the face of the earth, and like that is mapped down and designated by names. As these features remain unchanged, it is known that the moon always has the same hemisphere to us. Some of the cavities and basins are extremely bright, and have the power of reflecting light more brilliantly than others, just as the mountains and deserts of the earth are more reflective than the meadows and valleys. A green color prevails in many parts of the moon, probably the color of the rock. The height of the mountains, and the depth of the hollows, are known by the shadows cast when the sun is rising or setting on them. Before the sun's rays reach the general surface it lights up these peaks precisely as on earth. These mountains are less elevated than our highest, but are higher in proportion to the moon's size, some of them being about 3 or 4 miles high.

It has been asserted that works of art have been and may be seen in the moon. The improbability of this may be seen from these facts. The smallest portion of the moon's surface perceptible by the naked eye is equal to about seventy lunar miles; a distance of a mile in the moon subtends at the eye only an angle of one second.

The question of the moon's atmosphere has been much discussed, and there is much contradictory evidence as to its existence. If there is one it is probably very rare. There is good evidence that there is no water in the moon, for the edge of her crescent is always more or less jagged, whereas the presence of seas or lakes would make it partially smooth. No clouds have ever been discovered.

§ **501.** Since the axis of the moon is a degree and a half inclined to the axis of her revolution, and moreover as her revolution takes place in a plane inclined to the ecliptic, we sometimes see a little beyond one or the other pole. Sometimes, for the same reason, we see less of the illumined portion than we might expect. When she is full in the highest part of her orbit, a deficiency appears in the lower part of her disc, because we have not a full view of the enlightened hemisphere. When she is full in the lowest part of her orbit, there must be a similar deficiency observed in the upper edge. This is called the moon's libration in latitude.

She has also a libration in longitude, owing to her varying speed in her orbit, while her rotation is equable. In this way a strip a little west or east of the hemisphere usually presented is brought into view.

There is still another phenomenon of the same kind. The part of the moon presented to an observer at any place is bounded by a circle perpendicular to the line joining his place and the centre of the moon. To observers at different places, therefore, appearances in some degree different will be presented; for the moon is not so distant from the earth but that the lines joining her centre with different points on the earth's surface may make a sensible angle; in the extreme case, not less than twice the horizontal parallax of the moon, or nearly 2° on an average.

Every day presents another parallactic change in the moon's appearance, arising not from the different positions of the observers, but from the different positions of the moon as seen by an observer. When the moon rises in the east, an observer sees a little more of her western and then upper surface, than he would see if placed at the centre of the earth. When she is setting in the west, he sees a little more of her eastern and then upper side. This is called the diurnal libration.

Owing to all these causes combined, and also to a slight nutation of the moon's axis, we get sight of a zone a few degrees in breadth beyond an exact hemisphere of the moon.

§ **502.** The sun and stars rise and set to an inhabitant of the moon as they do to us, but only once a month instead of once in twenty-four hours. Since there is no atmosphere in the moon sufficiently dense to reflect light, the heavens in the day-time must have the appearance of night, and the stars must appear on a black ground, and as bright as they do in the night to us. Seen from the moon the earth must appear the largest body in the universe; its disc must be thirteen times as large as that of the moon seen from the earth. Since the moon's rotation would give the earth an apparent motion westward precisely equal to that which its revolution would give it eastward, the earth with regard to the moon will appear to stand still. It will be always visible in the same part of the heavens, though not among the same stars. Its only motion will be a slight apparent balancing caused by the libration of the moon.

It will always be invisible to one hemisphere of the moon and be continually seen from the other. Those to whom it is visible will see it exhibit all the phases which the moon presents to us, only at different times. The light received from the earth must prevent that hemisphere of the moon which is turned toward it from ever being in total darkness. It has alternately a fortnight of earth-light, and a fortnight of sun-light. But the other hemisphere of the moon has a fortnight's sun-light and a fortnight's darkness alternately. Hence its inhabitants, if there be any, can never see the earth unless they travel to gratify their curiosity.

While the earth turns on its axis the aspect it presents to the moon must be very various. Our seas, continents, forests and islands, must appear as so many spots of various brilliancy, and the atmosphere with its clouds must give still greater variety.

§ **503.** The light reflected from the earth to the moon must be very considerable since it is quite perceptible when again reflected back to us. The appearance called the new moon in the old moon's arms is the opaque body of the moon made visible by the light sent from the earth. It is seen only when the crescent of the moon is small. As the illumined portion of the moon seen by us increases it overpowers this ashy light, and the earth waning at the same time actually sends it less light. It has been thought that the moon possesses an innate light, but it is more probable that this light is refracted by the earth's atmosphere, and the state of the earth's atmosphere causes it to assume different hues in different eclipses. In some eclipses the moon retains almost all her light, usually however appearing of a red coppery hue. This may be owing to electrified vapors belonging to the earth's atmosphere, and interposed between it and the moon. Instances are recorded however where this feeble light has been entirely absorbed, so that the moon has altogether disappeared in her eclipses. It is not however certain that all the light of the moon when eclipsed comes from the earth.

§ **504.** The monthly phases of the moon are caused by her changes of position with regard to the earth. One half of her surface is always illumined, and as she revolves round us we see a larger or less portion of her illuminated surface. When her light is greatest she is most distant from the sun, and therefore upon the horizon longer than at any other time. When her light is least, when she shows but a portion of her illuminated surface, she remains visible above the horizon but a few hours after sunset or before sunrise. She is of course above the horizon as many hours on the average one solar day as another, but when she is full all these hours are night hours, when she is waxing or waning a portion of them are hours of daylight, and she is scarcely noticed except when pretty large in a pale wintry sky. When the moon is on the meridian at midnight her disc is entirely luminous, she is round and brilliant, she becomes visible when the sun sets, and sets when he rises. In a few days the bright part of her disc diminishes in breadth on the side farthest from the earth, she rises later and sets after sunrise. When she reaches her quarter her disc is reduced one half, she appears only the latter half of the night. Then she becomes a crescent whose horns are turned toward the west away from the sun, she rises later and later, the crescent wanes, the moon is dark; she rises with the sun and is seen no longer. She is usually invisible for several days, but the duration of her invisibility depends partly on climate, atmosphere, and power of vision.

§ **505.** An instance is on record of a lady, in the proverbially dingy atmosphere of England, noticing the old moon near her conjunction with the sun in the morning exhibiting a thread-like crescent, and the day after in the evening she observed the crescent turned the opposite way and eastward of the sun soon after sunset. Thus the same person saw, on the morning of one day and the evening of the next, a waning and a waxing moon. In Smyrna, when the atmosphere is exceedingly clear, the whole round dark blue disc of the moon is visible at the time of conjunction.

When the new moon appears east of the sun the horns of her crescent are turned from him, she remains above the horizon but a few hours. The crescent gradually increases, she follows the sun at a greater distance, becomes visible when he sets, and remains visible longer. She comes to her quarter, and still increasing, and remaining longer visible, at length comes in a line with the earth and sun, and is again a full moon. In whatever part of her monthly course the moon may be, and at whatever inclination to the ecliptic, if a line be imagined joining her horns, and bisected by another line at right angles to this, and produced beyond the convex part of the moon, the latter line will be in the direction of the sun.

§ **506.** The two points in the orbit corresponding to the new and full moon respectively are called by the common name of syzygies ; those which are 90° from the sun are called quadratures. The full moon is always in opposition to the sun, and consequently the full moons of winter are as much elevated above the equator as the sun is sunk below it, or as the sun is elevated above it in summer, and reciprocally it remains no longer above the horizon in summer than the sun does in winter. Thus all latitudes beyond the tropics have full moons of great altitude and which remain many hours visible in winter, but in summer their moons describe a lower and a shorter course. All that was formerly said of the sun's appearance to different portions of the earth applies to the moon also. These are the two full moons which occur about the tropics, all the others have a rising and a setting. During the six months' day of the poles, the full moon in the opposite part of the ecliptic is always invisible ; during their six months' night, the moon is visible from quadrature to quadrature, circling round the horizon for a fortnight at a time.

§ **507.** The moon's motion among the stars is so rapid as to be apparent in the course of a few hours. In the course of twenty-four hours she advances nearly 13°, and therefore to an observer on the earth rises later every day than the day before. If her revolution and our rotation were performed in the plane of the ecliptic, the moon would rise about three quarters of an hour later each day in the year than on the day preceding. But the moon's path is inclined to the equator, which causes unequal portions of it to rise in equal times, and it describes its course with a varying rapidity. These causes, however, make but a slight difference in the interval between two successive risings of the moon, and accordingly, to places on the equator, the moon rises about fifty minutes later each day than the preceding day. Within the tropics, there is so little variety of seasons that no one time is peculiarly harvest time. In higher latitudes, when the whole harvest of the year is gathered at once, the autumnal full moons give the husbandman an opportunity to complete his labors. The full moon of September rises soon after sunset for several evenings together, and is called the harvest moon. The full moon of October also rises not long after sunset, and nearly at the same time for some nights; it is called the hunter's moon. At the polar circles the autumnal full moon rises at sunset during the second and third quarters. At the poles the winter moons shine without setting during the same quarters.

§ **508.** It is not very easy to imagine the manner in which the moon rises at different angles to the horizon of places in high latitudes, but it may be seen on a globe. The moon's motion is so nearly in the ecliptic that we may consider her as moving in it. Now the different parts of the ecliptic, on account of its obliquity to the earth's axis, make very different angles with the horizon as they rise and set. Those parts or signs which rise with the smallest set with the greatest angles, and vice versa. When this angle is least a greater portion of the ecliptic rises in equal times, than when the angle is larger; as may be seen by elevating a globe to any considerable latitude, and turning it round on its axis. Consequently when the moon is in those signs which rise or set with the smallest angles, she rises or sets with the least difference of time; and with the greatest difference in those signs which rise or set with the greatest angles.

In northern latitudes, the smallest angle made by the ecliptic and horizon is when Aries rises, at which time Libra sets; the greatest when Libra rises, at which time Aries sets. Therefore the ecliptic rises fastest about Aries, and slowest about Libra. On the parallel of London as much of the ecliptic rises about Pisces and Aries in two hours as the moon goes through in six days; and therefore while the moon is in these signs, she differs but two hours in rising for six days together; that is, about twenty minutes later every day or night in a mean rate. But in fourteen days afterward the moon comes to Virgo and Libra, which are the opposite signs to Pisces and Aries; and then she differs almost four times as much in rising; namely, one hour and about fifteen minutes.

§ **509.** All these facts may be seen at once, by elevating the north pole of a globe to any desired altitude, and making chalk marks on the ecliptic at intervals of 13°, to represent the moon's mean place from day to day.

Then turning the globe westward seven of the marks about Pisces and Aries will rise in two hours and a half, measured by the motion of the index of the hour circle; but about the opposite signs the index will go over eight hours in the time that seven marks will rise. The intermediate signs will more or less partake of these differences as they are more or less remote from these signs.

Since the rising of the harvest moon is one of the few phenomena which may be better understood from a globe than by consideration of real events, I will mention one other mode of showing it. Let two celestial meridians on a celestial globe represent the edge of the illuminated hemisphere, and suppose the observer stationed on the brazen meridian; whenever the illumined edge comes under the brazen meridian, the observer will be 90° from the moon, and will see her rise. But this edge, as the globe turns round, will meet the observer at very varying intervals, and the nearer the observer is to the polar circle the greater will be the differences between these intervals.

§ **510.** The moon goes round the ecliptic in 27 days, 8 hours; but not from change to change in less than 29 days, 12 hours; so that she is in Pisces and Aries at least once in every lunation, and in some lunations twice. For while the moon goes round the ecliptic, from any conjunction or opposition, the earth goes almost a sign forward; and therefore the sun appears to go as far forward, that is $27\frac{1}{2}°$; so that the moon must go 27° more than round, and as much farther as the sun advances in that interval, which is $2\frac{1}{15}°$, before she can again be in conjunction with or opposition to the sun. Hence there can be but one conjunction or opposition of the sun and moon in a year in any particular part of the ecliptic. In the same way the hour and minute hands of a watch are never in conjunction or opposition in that part of the dial-plate where they were so last before.

§ **511.** As the moon can never be full but when she is opposite to the sun, and the sun is never in Virgo and Libra but in our autumnal months, it is plain that the moon is never full in the opposite signs, Pisces and Aries, but in these two months. And therefore we can only have two full moons in the year, which rise so near the time of sunset for a week together, as above mentioned.

Here it will probably be asked, why we never observe this remarkable rising of the moon but in harvest, seeing she is in Pisces and Aries twelve times in the year besides; and must then rise with as little difference of time as in harvest? The answer is plain: for in winter these signs rise at noon; and being then only a quarter of a circle distant from the sun, the moon in them is in her first quarter: but when the sun is above the horizon, the moon's rising is neither perceived nor regarded. In spring these signs rise with the sun, because he is then in them; and as the moon changes in them at that time of the year, she is quite invisible. In summer they rise about midnight, and the sun being then three signs, or a quarter of a circle, before them, the moon is in them about her third quarter; when rising so late, and giving but very little light, her rising passes unobserved. And in autumn these signs, being opposite to the sun, rise when he sets, with the moon in opposition, or at the full, which makes her rising very conspicuous.

§ **512.** In northern latitudes, the autumnal full moons are in Pisces and Aries; and the vernal full moons in Virgo and Libra: in southern latitudes, just the reverse, because the seasons are contrary. But Virgo and Libra rise at as small angles with the horizon in southern latitudes, as Pisces and Aries do in the northern; and therefore the harvest moons are just as regular on one side of the equator as on the other.

The moon's oblique motion with regard to the ecliptic causes some difference in the times of her rising and sitting from what is already mentioned. For when she is northward of the ecliptic, she rises sooner and sets later than if she moved in the ecliptic; and when she is southward of the ecliptic, she rises later and sets sooner. This difference is variable, even in the same signs, because the nodes shift backward about $19\frac{2}{3}°$ in the ecliptic every year; and so go round it contrary to the order of signs in 18 years, 225 days.

As there is a complete revolution of the nodes in 19 years, there must be a regular period of all the varieties which can happen in the rising and setting of the moon during that time. But this shifting of the nodes never affects the moon's rising so much, even in her quickest descending latitude, as not to allow us still the benefit of her rising nearer the time of sunset for a few days together about the full in harvest, than when she is full at any other time of the year.

§ **513.** Superstition has attributed to the moon great influence on chemical processes,—on the growth of seeds according as they are sown in the waxing or the waning moon,—on the weather, and the health and spirits of mankind. Many of the supposed effects are doubtless imaginary, some are not yet ascertained. Moon beams contain chemical rays and rays of heat; they have an effect similar but weaker than the sun's rays in daguerreotypes, so that it has been proposed to make the moon daguerreotype her own portrait. Sir John Herschel has often been quoted as believing that the moon influences the weather. This is all the influence he allows her. He thinks that the moon when at the full and a few days after, must be in a small degree a source of heat to the earth. But this heat emanating from a body below the temperature of ignition, will never reach the earth's surface, but will be arrested and absorbed in the upper strata of the atmosphere, where its whole power will be expended in converting visible cloud to transparent vapor. The rapid dissipation of clouds in moderate weather soon after the appearance of a full or nearly full moon, which he had himself observed on so many occasions, could, he thought, be explained on no other principle.

* § **514.** Several circumstances occur to render the moon's motions the most interesting, and at the same time the most difficult to investigate, of all the bodies of our system. In the solar system, planet disturbs planet; but in the lunar theory, the sun is the great disturbing cause;

* The rest of this chapter is taken from Mrs. Somerville's Connection of the Physical Sciences.

his vast distance being compensated by his enormous magnitude, so that the motions of the moon are more irregular than those of the planets; and, on account of the great ellipticity of her orbit, and the size of the sun, the approximations to her motions are tedious and difficult, beyond what those who are unaccustomed to such investigations could imagine. The moon is about four hundred times nearer to the earth than to the sun. The proximity of the moon to the earth keeps them together. For so great is the attraction of the sun, that if the moon were farther from the earth, she would leave it altogether, and would revolve as an independent planet about the sun.

§ **515.** The disturbing action of the sun on the moon is equivalent to three forces. The first, acting in the direction of the line joining the moon and earth, increases or diminishes her gravity to the earth. The second, acting in the direction of a tangent to her orbit, disturbs her motion in longitude. And the third, acting perpendicularly to the plane of her orbit, disturbs her motion in latitude; that is, it brings her nearer to, or removes her farther from the plane of the ecliptic than she would otherwise be. The periodic perturbations in the moon, arising from these forces, are perfectly similar to the periodic perturbations of the planets. But they are much greater and more numerous; because the sun is so large, that many inequalities which are quite insensible in the motions of the planets, are of great magnitude in those of the moon.

§ **516.** Among the innumerable periodic inequalities to which the moon's motion in longitude is liable, the most remarkable are, the equation of the centre, which is the difference between the moon's mean and true longitude, the evection, the variation, and the annual equation. The disturbing force which acts in the line joining the moon and earth produces the evection: it diminishes the eccentricity of the lunar orbit in conjunction and opposition, thereby making it more circular, and augments it in quadrature, which consequently renders it more elliptical. The period of this inequality is less than thirty-two days. Were the increase and diminution always the same, the evection would only depend upon the distance of the moon from the sun; but its absolute value also varies with her distance from the perigee of her orbit.

§ **517.** Ancient astronomers, who observed the moon solely with a view to the prediction of eclipses, which can only happen in conjunction and opposition, where the excentricity is diminished by the evection, assigned too small a value to the ellipticity of her orbit. The evection was discovered about A. D. 140. The variation produced by the tangential disturbing force, which is at its maximum when the moon is 45° distant from the sun, vanishes when that distance amounts to a quadrant, and also when the moon is in conjunction and opposition; consequently that inequality never could have been discovered from the eclipses; its period is half a lunar month. The annual equation depends upon the sun's distance from the earth; it arises from the moon's motion being accelerated, when that of the earth is retarded, and vice versa; for when the earth is in its perihelion, the lunar orbit is enlarged by the action of the sun; therefore the moon requires more time to perform her revolution. But as the earth approaches its aphelion, the moon's orbit contracts, and less time is necessary to accomplish her motion—its period, consequently, depends upon the time of the year. In the eclipses the annual equation combines with the equation of the centre of the terrestrial orbit, so that ancient astronomers imagined the earth's orbit to have a greater eccentricity than modern astronomers assign to it.

§ **518.** The planets disturb the motions of the moon both directly and indirectly; their action on the earth alters its relative position with regard to the sun and moon, and occasions inequalities in the moon's motion, which are more considerable than those arising from their direct action; for the same reason the moon, by disturbing the earth, indirectly disturbs her own motion. Neither the eccentricity of the lunar orbit, nor its mean inclination to the plane of the ecliptic, have experienced any changes from secular inequalities; for, although the mean action of the sun on the moon depends upon the inclination of the lunar orbit to the ecliptic, and the position of the ecliptic is subject to a secular inequality, yet analysis shows, that it does not occasion a secular variation in the inclination of the lunar orbit, because the action of the sun constantly brings the moon's orbit to the same inclination to the ecliptic.

The mean motion, the nodes, and the perigee, however, are subject to very remarkable variations.

§ **519.** From the eclipse observed by the Chaldeans at Babylon, on the 19th of March, 721 years before the Christian era, the place of the moon is known from that of the sun at the moment of opposition, whence her mean longitude may be fouud. But the comparison of this mean longitude with another mean longitude, computed back for the instant of the eclipse from modern observations, shows that the moon performs her revolution round the earth more rapidily and in a shorter time now than she did formerly, and that the acceleration in her mean motion has been increasing from age to age as the square of the time. All ancient and intermediate eclipses confirm this result. As the mean motions of the planets have no secular inequalities, this seemed to be an unaccountable anomaly. It was at one time attributed to the resistance of an ethereal medium pervading space, and at another to the successive transmission of the gravitating force. But as La Place proved that neither of these causes, even if they exist, have any influence on the motions of the lunar perigee or nodes, they could not affect the mean motion; a variation in the mean motion from such causes being inseparably connected with variations in the motions of the perigee and nodes. He perceived that the secular variation in the elements of Jupiter's orbit, from the action of the planets, occasions corresponding changes in the motions of the satellites, which led him to suspect that the acceleration in the mean motion of the moon might be connected with the secular variation in the eccentricity of the terrestrial orbit. Analysis has shewn that he assigned the true cause of the acceleration.

§ **520.** It is proved that the greater the eccentricity of the terrestrial orbit, the greater is the disturbing action of the sun on the moon. Now as the eccentricity has been decreasing for ages, the effect of the sun in disturbing the moon has been diminishing during that time. Consequently the attraction of the earth has had a more and more powerful effect on the moon, and has been continually diminishing the size of the lunar orbit. So that the moon's velocity has been gradually augmenting for many centuries to balance the increase of the earth's attraction. This secular increase in the moon's velocity is called the acceleration, a name peculiarly appropriate at present, and which will continue to be so for a vast number of ages ; because, as long as the earth's eccentricity diminishes, the moon's mean motion will be accelerated ; but when the eccentricity has passed its minimum, and begins to increase, the mean motion will be retarded from age to age. The secular acceleration is now about 11″.9, but its effect on the moon's place increases as the square of the time. It is remarkable that the action of the planets, thus reflected by the sun to the moon, is much more sensible than their direct action either on the earth or moon. The secular diminution in the eccentricity, which has not altered the equation of the centre of the sun by eight minutes since the earliest recorded eclipses, has produced a variation of about 1° 48′ in the moon's longitude, and of 7° 12′ in her mean anomaly.

§ **521.** The action of the sun occasions a rapid but variable motion in the nodes and perigee of the lunar orbit. Though the nodes recede during the greater part of the moon's revolution, and advance during the smaller, they perform their sidereal revolution in 6,793d. 9h. 23′ 9″.3 ; and the perigee accomplishes a revolution in 3,232d. 13h. 48′ 29″.6, or a little more than nine years, notwithstanding its motion is sometimes retrograde and sometimes direct : but such is the difference between the disturbing action of the sun and that of all the planets put together, that it requires no less than 109,830 years for the greater axis of the terrestrial orbit to do the same, moving at the rate of 11″.8 annually. The form of the earth has no sensible effect either on the lunar nodes or apsides. It is evident that the same secular variation which changes the sun's distance from the earth, and occasions the acceleration of the moon's mean motion, must affect the nodes and perigee. It consequently appears, from theory as well as observation, that both these elements are subject to a secular inequality, arising from the variation in the eccentricity of the earth's orbit, which connects them with the acceleration, so that both are retarded when the mean motion is anticipated.

§ **522.** The moon is so near that the excess of matter at the earth's equator occasions periodic variations in her longitude, and also that remarkable inequality in her latitude, already mentioned as a nutation in the lunar orbit, which diminishes its inclination to the ecliptic when the moon's ascending node coincides with the equinox of spring, and augments it when that node coincides with the equinox of autumn. As the cause must be proportional to the effect, a comparison of these inequalities, computed from theory, with the same given by observation, shows that the compression of the terrestrial spheroid, or the ratio of the difference between the polar and equatorial diameters, to the diameter of the equator, is $\frac{1}{305}$. It is proved analytically, that if a fluid mass of homogeneous matter, whose particles attract each other inversely as the squares of the distance, were to revolve about an axis as the earth does, it would assume the form of a spheroid whose compression is $\frac{1}{230}$. Since that is not the case, the earth cannot be homogeneous, but must decrease in density from its centre to its circumference. Thus the moon's eclipses show the earth to be round; and her inequalities determine not only the form, but also the internal structure of our planet; results of analysis which could not have been anticipated. Similar inequalities in the motions of Jupiter's satellites prove that his mass is not homogeneous, and that his compression is $\frac{1}{13.8}$.

CHAPTER XXIII.

ECLIPSES.

Conditions necessary for an Eclipse. Lunar Eclipses. Dimensions of the Earth's Shadow. Limits of a Lunar Eclipse. Solar Eclipse. Effect of the Moon's Parallax. Limits of a Solar Eclipse Number of Eclipses in a year. Eclipse of 1706. Eclipse of 1842.

§ **523.** Since the sun and moon are equally concerned in eclipses, their details have been reserved till we were fully acquainted with the motions of both these bodies. To make an eclipse three bodies are necessary,—a light-giving, a light-receiving, and a light-intercepting body. These three bodies must be wholly or partially in one straight line. If the moon is new and intercepts the light, we have a solar eclipse ; if the moon is full and the earth cuts off the light, we have a lunar eclipse. In the last case the moon is actually deprived of light by passing into the earth's shadow ; in the first case only a small portion of the earth is really deprived of light or eclipsed, but to that portion the whole or part of the sun appears eclipsed, though in fact it is as bright as ever. When the moon is eclipsed, the sun appears eclipsed to her, totally so to all those parts on which the earth's shadow falls, so long as they are in the shadow. When the sun is eclipsed to us, the moon's inhabitants, if she has any, see her shadow like a dark spot travelling over the earth, about twice as fast as its equatorial parts move, and in the same direction.

§ **524.** In order that either kind of an eclipse should happen, one body must pass into the shadow of another or into the line of direction of its shadow. Every planet, and every satellite, casts a shadow toward that point of the heavens which is opposite the sun. We might therefore expect eclipses to occur continually. But the primaries are too distant from one another ever to enter one another's shadow; the only eclipses which can take place are between the primaries and secondaries which are so much nearer. The occurrence of eclipses is made more frequent both by the greater length and the greater circumference of a shadow. A shadow may be cast by a large body, but if the one which gives light be vastly larger the shadow will be a short cone. If the light-giving and light-intercepting bodies were of the same size, the shadow would be a cylinder extending infinitely; in this case the body would be equally eclipsed at whatever distance from the intervening body it passed. If the shadow were a cone, the body would be eclipsed longer if it passed near the base of the cone. Since the sun is so much larger than the planets, their shadows are all conical and not very long. The shadow of the earth where the moon passes through it is large enough to cover the moon's disc if its diameter were three times what it now is; the moon's shadow often ends before it can reach the earth, and never, unless it falls very obliquely, covers more than 170 linear miles of its surface.

If the moon's orbit coincided with the plane of the ecliptic there would be a lunar eclipse every full moon, and a solar eclipse every new moon; the moon would be eclipsed to all one hemisphere of the earth for an hour and a half; the sun would to certain small portions of the earth appear darkened for a space not exceeding eight minutes. Since the moon's orbit is inclined, eclipses only take place when the moon is new or full near her nodes.

§ **525.** We shall begin with lunar eclipses, because they are more simple. In considering them we need not refer to any particular point on the surface of the earth, or embarrass ourselves with observations made in different places, for her disappearance is absolute and universal; we have only to ascertain when the light transmitted from the sun actually ceases to reach the moon.

If the moon happens to be full in her node, the centre of her disc will pass through the centre of the earth's shadow; the eclipse will be total and of the longest duration. If the moon is full near her nodes, she may enter one or the other edge of the earth's shadow, and there may be a partial eclipse for a shorter time. If she is full 90° from her node, her centre will be upwards of 5° from the centre of her shadow, and there will be no eclipse. There must be a certain distance from her node, beyond which there can be no eclipse. This is called the lunar eclipse limit. It may be found by calculating the size of the earth's shadow at the mean distance of the moon from the earth, and adding to its radius the radius of the moon's disc. If the right line joining the ecliptic and the moon's orbit at a given distance from the node be less than these, there may be an eclipse ; if more, no eclipse is possible. Where the distance from the moon's orbit to the ecliptic equals these two quantities added, is the lunar eclipse limit.

§ **526.** The distance of the earth from the sun and the sun's and the earth's sizes being known, the length of the earth's cone may be found by similar triangles. It extends 220 of the earth's radii, while the moon is only 60 of such radii distant. The breadth of the shadow at the mean distance of the moon may be found by the following proportion. As 220 radii : 220—60 : : 1 : $\frac{8}{11}$ of the earth's diameter, or of 7,912 miles. As the diameter of the moon is 2,160 miles, and as the angles are small, we may find the angle this shadow will subtend by the proportion, 2,160 : 5,754 : : the angular diameter of the moon : to the angular diameter of the earth's shadow at the distance of the moon.

The farther the sun is from the earth the more slowly do the boundaries of the shadow approach each other, and the larger therefore is the shadow at the moon's distance. The nearer the moon is to the earth, the larger, other things continuing the same, is the part of the shadow through which she passes. On both accounts the duration of an eclipse is greatest when the moon is at the least, and the sun at the greatest distance. These causes also make some difference in the limits within which an eclipse can take place.

§ **527.** Taking the extreme values, the greatest apparent radius of the shadow is 45′ 12″.15, and the corresponding apparent radius of the moon is 16′ 5″.45 ; and the greatest distance from the centre of the earth's shadow, at which the moon can possibly come in contact with it, is the sum of these quantities, or 62′ 37″.65. In the same manner the least apparent radius of the shadow is 36′ 42″.15, and the correspondent apparent radius of the moon is 14′ 41″, and the least distance at which the moon can just be in contact with the shadow and no more, is the sum of these quantities, or 51′ 23″.15. If therefore the moon never comes so near as 62′ 37″.65 to the centre of the earth's shadow, there can be no eclipse; if she comes to that distance or within it, there may; if she comes within the distance of 51′ 23″.15, there must be an eclipse.

Let N M (Fig. 14, Plate II.) represent a portion of the moon's orbit, N E a portion of the ecliptic, N of course being the node. Let E M, a secondary to M n, be 62′ 37″. If we can ascertain what must be the value of E N to correspond with this value of E M, we shall ascertain how distant the node may be from the centre of the earth's shadow to admit of there being an eclipse of the moon. The angle E M n is a right angle, E N M is 5° 17′, and the side E M is known by supposition. The remaining sides and angles may therefore be computed. E N is equal to 11° 25′ 40″ nearly. If E M is taken=51′ 23″, E N will equal 9° 20′ 29″ nearly. An eclipse may or must take place within these limits on each side of the node.

§ **528.** We have hitherto spoken of the shadow as conical; and it is true that the portion of space within which the earth will entirely conceal the sun is so. This conical shadow is called the umbra. But there will be another portion within which a part of the sun will be concealed. Beyond the umbra are her diverging spaces, where if a spectator be situated he will see only a portion of the sun's surface, the rest being obscured by the earth. B C and A D, two common tangents to the sun and the earth, drawn crossing the line which joins their centres, give the limits of this faint shadow or penumbra on both sides. The penumbra lies on all sides of the umbra.

In a lunar eclipse, the moon enters the penumbra first, and gradually gets involved in the umbra. It is difficult to ascertain the moment of passing from one to the other, and for this reason eclipses of the moon cannot give terrestrial longitude exactly. When the centre of the moon passes through the centre of the shadow, the eclipse is called central or total. When the moon passes through the upper or lower portion of the shadow, the eclipse is called partial. In order to mark the extent of the eclipse, the diameter of the moon (or sun) is supposed to be divided into twelve equal parts, called digits, and the depth of the immersion is estimated in digits. When the moon enters the penumbra only, it is not said to be eclipsed.

§ **529.** The consideration of a solar eclipse is more embarrassing. One calculation for the whole earth will not answer here; the position of the spectator on the earth's surface, and even the rotation of the earth must be allowed for; and the duration and extent of the eclipse must be computed for particular places. The sun is seen from the earth in nearly its true place, but the moon's parallax is considerable, and must not be neglected in finding the limits within which a solar eclipse may take place. By parallax the moon's apparent edge may be thrown in any direction according to the spectator's station, by any amount not exceeding the horizontal parallax. Now this comes to the same, so far as the possibility of an eclipse is concerned, as if the apparent diameter of the moon, seen from the earth's centre, were dilated by twice its horizontal parallax; for if when so dilated it can touch or over lap the sun, there must be an eclipse at some part or other of the earth's surface. This sum is at its maximum about 1° 34′ 27″. Trom this the lunar ecliptic limit is found to be 17° 21′ 27″, when the sun is farthest and the moon nearest; 15° 14′ 27″, when the sun is at its least and the moon at its greatest distance, and therefore appears smallest. If then at the moment of the new moon the moon's node is farther from the sun in longitude than this limit, there can be no eclipse; if within 17°, there may; and if within 15°, there must be one, to some part of the earth. To ascertain for any place whether there will be one or not, and also its extent, the place of the node and the semi-diameters must be exactly ascertained, and the local parallax, and also the increase in the moon's apparent diameter, owing to the spectator's being nearer her than the centre of the earth is, must be found.

§ **530.** When the moon is in perigee while the sun is in apogee, her distance from the centre of the earth is not quite sixty radii of the earth, her shadow reaches to the earth, and to some portion of the earth there will be a total eclipse. When the moon is in apogee while the sun is in perigee, her distance is nearly sixty-four radii, the earth is beyond the termination of the shadow, and there can be no total eclipse, though there may be an annular one, or one in which the moon covers the central part of the sun's disc and leaves a ring-shaped surface visible. Intermediate positions of the sun and moon require corresponding calculations; in general when the apparent diameter of the moon exceeds that of the sun, the eclipse is total, when the sun's is the largest, it is annular.

The portion of the earth's surface at which the eclipse is total at the same moment cannot exceed a circle with a radius of eighty-eight miles, if the centre of the earth is in a line with the axis of the shadow. If the centre of the earth be not in this line, the radius of the shadow will be less, but the shadow will fall more obliquely.

The whole region of the earth, however, to which the eclipse may be total is greater than has been stated. The motion of the moon carries her shadow along a zone or belt of the earth's surface, and the eclipse is total, though at different times to the inhabitants of different parts within this belt. Thus the time at which a total eclipse takes place is different for different places. Its duration will also be different as the centre of the shadow, or only a more remote part of it, passes over the spot. It can never continue total at any particular place for more than 7′ 38″, nor be annular for more than 12′ 24″. An annular eclipse must be longer than a total one, because the sun's diameter being the largest the moon occupies more time in traversing it. In partial eclipses the observer is only within the penumbra of the moon, and more or less of the sun is hidden, as the observer is less or more remote from the centre of the sun. The penumbra covers a circle or ellipse of about 2,000 miles radius.

§ **531.** The duration of eclipses is modified by the motions of the moon and of the earth. Owing to the moon and earth's motion being round the sun in the same direction, lunar and solar eclipses are slightly lengthened, the moon keeps in the earth's shadow longer than if the earth stood still. The moon's revolution round the earth shortens by a minute quantity solar eclipses. The varying rapidity with which the earth and moon move in different parts of their orbits, introduces another cause of irregularity into eclipses. The greatest total eclipse is shortened because the moon is then in her perigee; annular eclipses are lengthened because the moon is then in apogee. The motion of the earth in rotation has no effect on lunar eclipses. Since the moon passes the same way, but twice as fast, a solar eclipse is lengthened a little to most places. But if the earth is so inclined that the eclipse extends over one of its poles, an observer in that region would be carried through the shadow more quickly than if he were carried by the earth's rotation.

§ **532.** As the solar eclipse limits exceed the lunar, there must be more eclipses of the sun than of the moon. But every eclipse of the moon is visible wherever the moon is above the horizon at the time when it takes place, that is to half the earth; and as she is above the horizon of each particular place as long during the year as she is below it, half of her eclipses are visible to each observer wherever situated. Not only do half of the eclipses of the sun take place while he is below the horizon of a certain place, but, as we have seen, it may not be visible at many places while he is above the horizon. Thus, though the whole number of solar eclipses exceeds that of the lunar, the number visible at any particular place falls short of it.

Since the solar eclipse limits are from 30° to 34°, there must be at least one new moon and perhaps two while the sun is so near the node as to be eclipsed. This may happen at both nodes in one year; and as the nodes retreat 19° in the course of a year, the sun may again come round so near the node as to be eclipsed. Hence there may be five solar eclipses in one year. Since the lunar eclipse limits do not exceed 22°, and the sun is less than a month in moving through that space, there may not be any full moon near the node, and consequently no lunar eclipse.

But if there are two solar eclipses at one node, there must be one lunar between them. And also if there be one lunar eclipse near a node there must be at least one solar eclipse also. The greatest number of eclipses which can take place in a year is seven; five solar, and two lunar. The least number is two, both of which will be solar. When there are four solar, there may, owing to the motion of the nodes, be three lunar; but when there are five solar, there can be but two lunar.

§ **533.** After a certain period eclipses return very nearly in the same order and of the same magnitude. 223 of the moon's synodical revolutions occupy 6585.32 days, and nineteen complete synodical revolutions of the node occupy 6585.78. After this period eclipses return very nearly in the same order as before, though not so accurately as to dispense with the trouble of calculating them. The period of eighteen years and ten days was early discovered by the Chaldeans and used by them in foretelling eclipses. Another period, called the Metonic cycle, consists of 235 synodical revolutions, or 19 tropical years. New and full moons fall on the same day in every Metonic cycle; it has therefore been much used in regulating games and feasts and fasts. By means of these cycles, and our knowledge of the laws of eclipses, the date of some historical events has been fixed with great precision. The date of the battle of Arbela has been determined from its being fought eleven days after an eclipse whose period has been computed. And a battle between the Medes and the Lydians, which was broken off in consequence of an eclipse, and of which even the year was not known, has been similarly investigated, for there was only one eclipse about that time which could be total in the part of Asia where it was fought.

§ **534.** Since so many circumstances are necessary to produce total or annular eclipses, their occurrence at any one place is extremely rare. The eclipse of 1706 was total for a long time over a great extent of country in Europe, from Seville, crossing Spain diagonally, the southern parts of France, part of Switzerland and Germany, Poland, and the countries to the northeast, even to the Frozen Ocean.

At Montpellier total darkness lasted nearly five minutes. The obscurity resembled neither real night nor twilight. Planets and stars were to be seen. The affrighted animals deserted their pastures, and sought their stables; birds of night left their retreats, and those of the day sought for shelter. Round the obscure disc of the moon was a luminous ring which becoming fainter, extended 4° 30′ on all sides. It was not however a ring of the solar disc, for the apparent diameter of the moon exceeded that of the sun by more than two minutes. The light also was much more pale and gloomy than that of the sun in an annular eclipse.

§ **535.** In 1842 another total eclipse of the sun, visible on the continent of Europe, occurred. Long journeys were undertaken by distinguished astronomers from the desire to view this rare phenomenon. Before leaving home they had agreed not to interchange one word on the subject of the eclipse till each had written and published his own observations. Each one, on his return, described in the same manner the striking peculiarities of this eclipse, the luminous ring, which must have been the sun's atmosphere, surrounding the dark disc of the moon, the bursting out of three large red protuberances or tongues of flame from the edge of the moon, evidently connected with this ring or crown of light, and the rainbow changes of color which these protuberances took on. These tongues of light remained unchanged in shape during the eclipse after they had once appeared, proving themselves to be realities and not an optical illusion. This eclipse was not only interesting from the great beauty of the corona and the rose-colored lights, but it seems also to prove the non-existence of an atmosphere in the moon, and the existence of one round the sun. Arago infers that the moon has no atmosphere from the perfectly well-defined horns of the sun's crescent. If the moon had ever so thin an atmosphere, the rays from these horns passing over the dark portion of the moon before they met the earth, would be deflected, and the outline of the crescent would be injured; but nothing of this kind was seen. The luminous ring did not belong to the moon, for it did not move with it; on the contrary, the moon appeared to glide in front of it and of the colored heights. The ring was described by one observer as one third, by another as one eighth part of the sun's diameter. Sir John Herschel considers the tongues, which were by different observers variously described as flaming or icy mountains, to be rose-colored clouds floating in the sun's atmosphere.

§ **536.** The effect of the eclipse upon the population of Perpignan, who were watching it, is described by M. Arago as singular and even affecting. The gravest persons were unable to restrain expressions of joy when the sun re-appeared ; and whilst the eclipse lasted, anxiety was depicted on every countenance. The effect upon animals was remarkable. One of the friends of M. Arago had placed five healthy linnets in a cage. During the sudden darkness of the eclipse, three of the five died. The oxen formed into a circle, with their horns thrust forward, as if preparing for the attack of an enemy. At Montpellier, bats and owls left their retreats, and sheep laid down as for the night, and the horses in the fields were in a state of terror. In addition to these facts, it is said, that a swarm of ants in full march stopped short at the moment of occultation.

At Paria, over which town the line of central darkness exactly passed, at the moment when the total obscuration commenced, a brilliant crown of glory encircled the moon, like the aureola, which painters append to saints. Suddenly, from the border of the black and laboring moon, thus singularly enshrined, burst forth at three distinct points, within the aureola, purple or blue flames, visible to every eye. At this moment, from the whole assembled population of the town, a simultaneous and deafening shout broke forth.

ERRATA.

Page 37, line 7, from the bottom, *dele* 'others.'
" 40, " 2, " *for* 'the,' *read* 'two.'
" 77, top line, *for* 'common,' *read* 'cannon.'
" 110, line 15, from the bottom, *for* 'eighty-third,' *read* 'sixty-seventh.'
" 110, " 13, " *for* 'eighty-three, *read* 'sixty-seven.'
" 111, " 22, from the top, *for* 'aphelion,' *read* 'perihelion.'
" 143, " 16, " *for* 'rate,' *read* 'ratio.'
" 143, " 12, from the bottom, *for* '$\sqrt{51}$,' *read* '$\sqrt{57}$.'
" 165, " 16, " *for* 'be,' *read* 'increase.'
" 165, " 11, " *for* 'actual,' *read* 'increase of.'
" 171, " 7, from the top, *for* 'raised,' *read* 'caused.'
" 171, " 24, " *for* 'south,' *read* 'north.'
" 171, " 26, " *for* 'north,' *read* 'south.'
" 172, " 18, " *for* 'Europe to America,' *read* 'America to Europe.'
" 204, *for* columns of Continual Day, insert these :

37	35
67	53
95	90
127	121
159	152
192	184

Dele columns of Continual Night.

" 261, line 5, *for* 'E,' *read* 'C.'
" 261, " 9, *for* '2 a p,' *read* '2 A p.'
" 262, " 11, *for* 'minute,' *read* 'second.'
" 263, " 5, *for* '$\sqrt{2}$ g r,' *read* '$\sqrt{}$g r.'
" 263, " 23, *for* 'minute,' *read* 'second.'
" 263, " 27, *for* 'As 27, 7′, 43″ : 1″ : : 360° : 33″,' *read* "27, 7, 43′ : 1″ : : 360 : $\frac{33''}{60}$.
" 265, " 7, from the bottom, *for* '3183," *read* '.3183.'
" 265, " 4, " *for* '3183^2, or 10131489,' *read* '$.3183^2$, or .10131489.'
" 308, lines 6, 5, 4, " *for* 'S,' *read* 'T.'
" 315, line 10, from the top, *for* 'west,' *read* 'east.'
" 334, " 19, " *for* 'm d^2 : M D^2,' *read* 'm D^2 : M d^2.'
" 353, " 5, " *for* 'lunar,' *read* 'linear.'
" 355, " 18, from the bottom, *for* 'it,' *read* 'the sun.'
" 355, " 7, " *for* 'upon,' *read* 'above.'
" 356, " 12, from the top, *insert after* 'also.' 'As the polar circles have each one day and one hour of 24 hours, they have also one full moon which remains 24 hours above, and another which remains as long below the horizon.
" 371, " 14, *for* 'From this the lunar,' *read* 'From this the solar.'

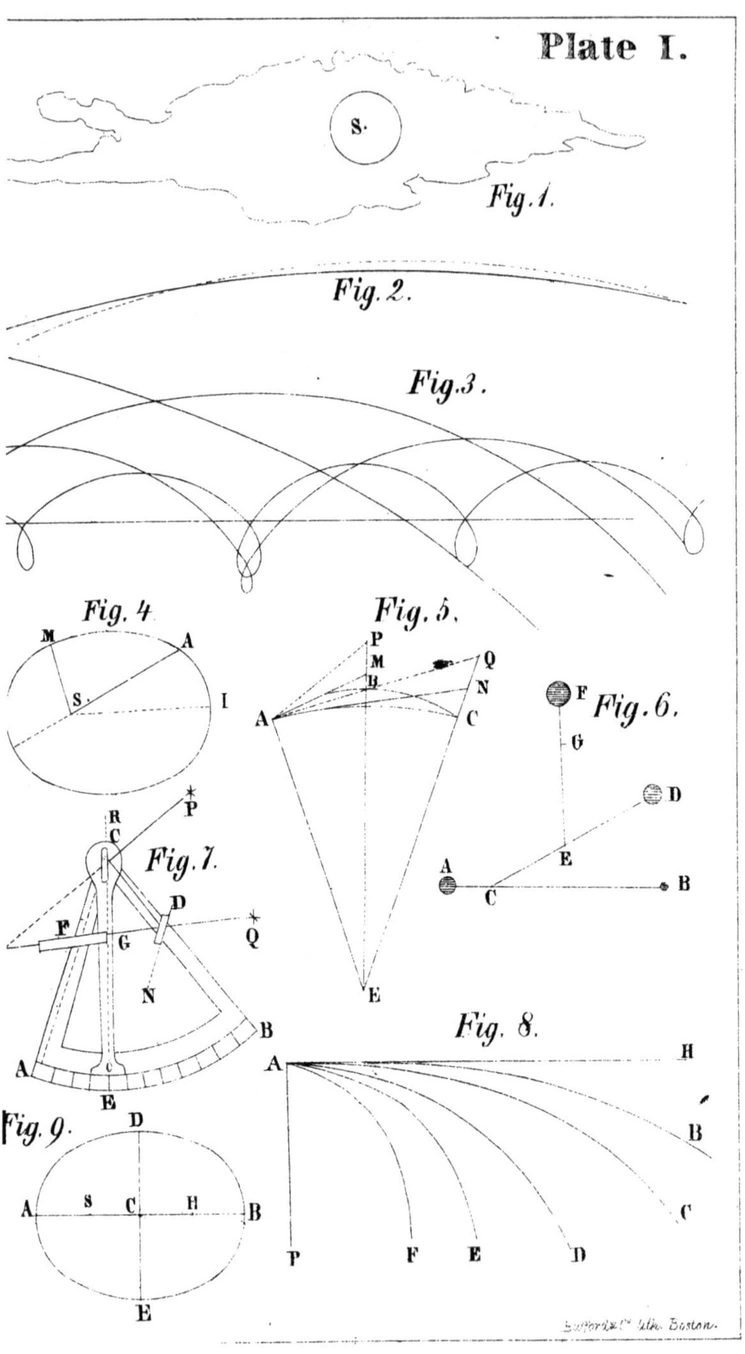

Plate I.
S.
Fig. 1.
Fig. 2.
Fig. 3.
Fig. 4
M
A
S.
I
Fig. 5.
P
M
Q
N
A
C
E
F
Fig. 6.
G
D
E
A
C
B
R
C
P
Fig. 7.
D
F
G
Q
N
B
A
E
Fig. 8.
A
H
B
C
P
F
E
D
Fig. 9.
D
A
S
C
H
B
E

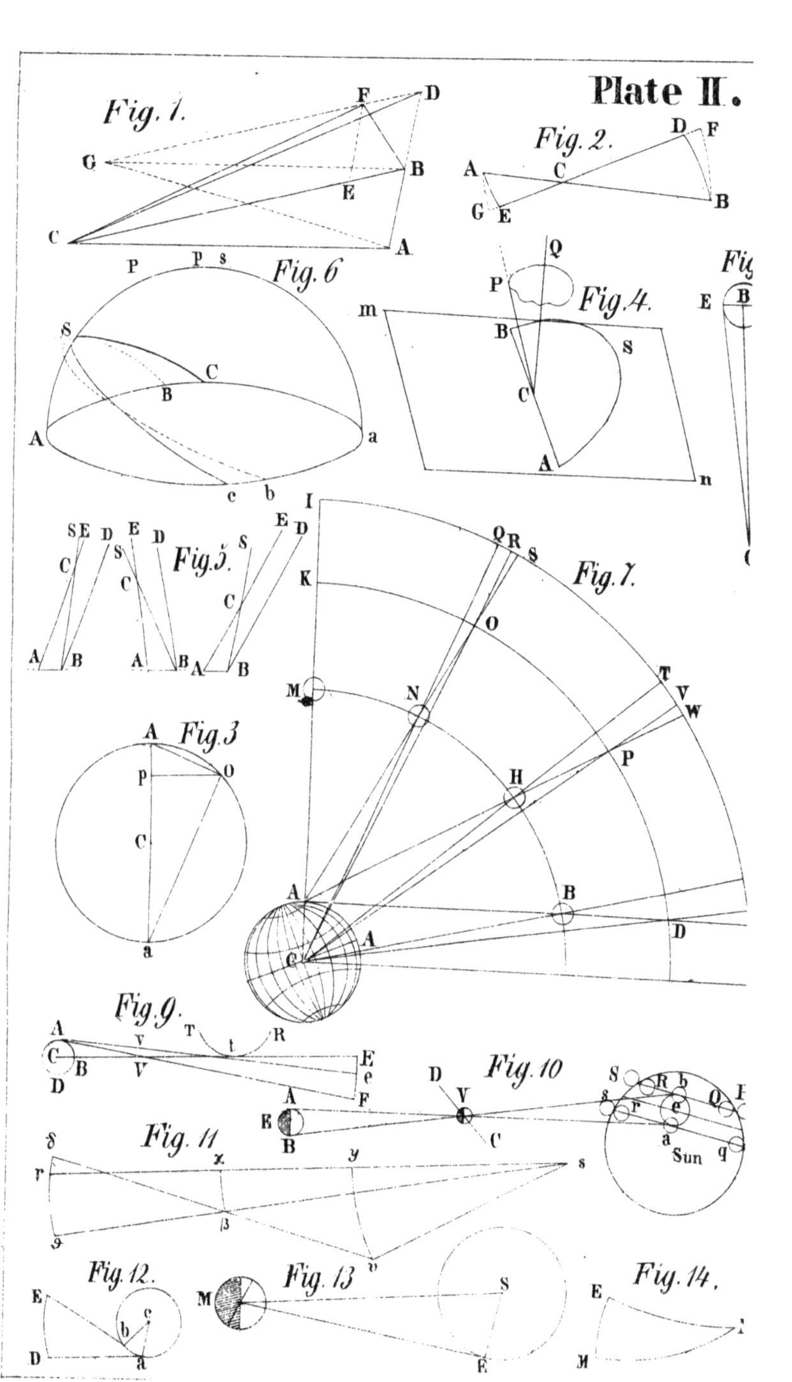
Plate II.
Fig. 1.
Fig. 2.
Fig. 3
Fig. 4.
Fig. 5.
Fig. 6
Fig. 7.
Fig. 9.
Fig. 10
Fig. 11
Fig. 12.
Fig. 13
Fig. 14.
Sun

www.ingramcontent.com/pod-product-compliance
Lightning Source LLC
LaVergne TN
LVHW010132110826
845151LV00002B/339

* 9 7 8 1 4 2 5 5 3 9 8 2 5 *